T0335551

Comparative Physiology, Natural Animal Models and Clinical Medicine

Insights into Clinical Medicine from Animal Adaptations

Comparative Physiology, Natural Animal Models and Clinical Medicine

Insights into Clinical Medicine
from Animal Adaptations

Michael A Singer
Faculty of Medicine, Queen's University, Canada

Imperial College Press

ICP

Published by

Imperial College Press
57 Shelton Street
Covent Garden
London WC2H 9HE

Distributed by

World Scientific Publishing Co. Pte. Ltd.
5 Toh Tuck Link, Singapore 596224
USA office: 27 Warren Street, Suite 401-402, Hackensack, NJ 07601
UK office: 57 Shelton Street, Covent Garden, London WC2H 9HE

British Library Cataloguing-in-Publication Data
A catalogue record for this book is available from the British Library.

ISBN-13 978-1-86094-782-7
ISBN-10 1-86094-782-4

Typeset by Stallion Press
Email: enquiries@stallionpress.com

Printed in Singapore.

*This book is dedicated to
the late Howard S. Frazier;
mentor, teacher and friend.*

Contents

Acknowledgements

I would like to thank my wife Bunny for support and encouragement throughout the book writing process. I would like to thank my colleagues, A. Ross Morton, Robert Kisilevsky and Leonard Finegold for reading selected chapters and Connie Latimer for secretarial help in preparing the manuscript.

Introduction

It is a winter day and Mr. Jones, a 45-five-year-old man suffering from chronic renal failure has just arrived at his regional dialysis center. He comes here three times a week, each time for four hours, to be connected to a hemodialysis machine. These treatments are necessary for Mr. Jones to stay alive. During each 4-hour treatment his entire blood volume will pass through the artificial kidney machine about 14 times for purification. On this same day, many miles away, an American black bear slumbers in its wintery cave. The bear will remain there dormant for up to five months during which time this animal will not eat, drink, defecate or urinate. Although dormant, the bear still has an active metabolic rate about 50% of normal. Yet despite having no urine output for this prolonged period of time, the bear will not suffer any of the manifestations of renal failure experienced by Mr. Jones. How has the bear's metabolic machinery adapted to such a prolonged state of functional renal failure? Can we learn new approaches for the prevention and/or treatment of chronic renal failure from such a natural animal model?

Natural selection is the mechanism underlying the process of evolution. Changing environmental conditions select out animals whose metabolic/physiologic characteristics confer on them a survival advantage. Evolution can be considered a natural experimental process in which over billions of years countless animal design features have been tested. In some cases, an animal group or species has evolved a set of biochemical/physiological features natural and adaptive for that animal but quite abnormal for humans. For example, many species of birds have extremely high blood glucose concentrations which are clearly natural and adaptive for these birds but would be considered quite abnormal and in fact in the diagnostic range for diabetes if found in humans. Despite these high blood glucose levels, these birds do not suffer vascular and renal complications as observed in humans with diabetes. Can the study of these avian models give us insights into the pathogenesis of diabetic complications in humans? Patients with hepatic disease frequently suffer from ammonia neurotoxicity. Fish on the other hand, are ammonia tolerant as compared to mammals. Can fish teach us how to prevent ammonia neurotoxicity? In this book, a natural animal model is defined as an animal group or species that possesses biochemical/physiological characteristics natural for that animal but pathological for humans. By using these models, we take advantage of all the "research" that nature has already performed in animal design testing and selection. The premise of this book is that by studying natural animal models we can gain valuable insights into the etiology and treatment of various clinical disorders.

Animals have been used extensively as subjects in medical research, but not generally in their natural state. One approach is to induce a disease in the animal either through pharmacologic or genetic techniques. Sometimes a disease occurs spontaneously in a few members of an animal group and then these animals can be bred to create a colony of disease-afflicted animals. With these models the implicit assumption is made that the disease is similar to that observed in humans. We can use such models to investigate pathogenic mechanisms underlying the disease and also to explore possible treatment options. Natural animal models do not suffer from a disease and hence provide a different perspective. With these models, we can examine biological solutions to clinical disorders that nature has tested through the cauldron of evolution and proven to be effective. Birds, which naturally have chronically high blood glucose levels, do not suffer from diabetes. The question we must ask of such a natural model is how do these birds tolerate blood glucose concentrations, which in the human are associated with pathological consequences. The natural model is living proof that a biological answer to this question is available.

Most biochemical/physiological processes are multifactorial. For example, consider the hepatic synthesis of urea in the mammal. Production of urea in the liver serves to detoxify ammonia liberated as a result of amino acid catabolism. In addition, urea is also an important component of systems involved in nitrogen and water conservation. Some of the produced urea is transferred across the gut wall where the nitrogen can be salvaged via the action of bacteria, and some of the urea is transferred across the wall of the kidney tubule into the surrounding tissue where its accumulation aids in water reabsorption. This concept of multifunctionality is important. If we were to consider urea only as a vehicle for ammonia detoxification, then this molecule might be considered a poor choice since it is metabolically expensive to synthesize. However, since this same molecule also functions in systems designed to conserve water and nitrogen then its "value" becomes enhanced considerably.

Multifunctionality is an example of the economy of animal design characteristics. If we were to consider a single animal system in isolation, we would probably conclude that it was not designed optimally in terms of the matching of structure and function. However, if we consider this same system in light of all the functions that it serves we would conclude that it was optimally designed to fulfill these multiple functions. The solutions to clinical disorders that natural animal models give us represent adaptations that have evolved within this context of multifunctionality of body systems. In contrast, consider the animal that has been induced to have a specific disease. We might design a drug to treat that disease based on studies done in such an animal model. This drug will probably have a number of unexpected side effects since the drug has only been designed to affect one of the functions of the targeted receptor or molecule thought to be involved in the disease process. Since multifunctionality is generally the rule, side effects will arise from the drug interfering with the other as yet unidentified functions of that receptor or molecule. On the other hand, the solutions that nature gives us will not have side effects since they would have evolved within a multifunctionality framework.

We might consider that a bird and a turtle are so different from humans that such natural animal models have no relevance to clinical disorders. However, the cornerstone of Darwin's theory is that all organisms had common ancestors and that probably all life on Earth started from a single origin of life. Some basic genes of higher organisms can be traced all the way back to homologous genes in bacteria. In addition, despite the incredible diversity of animals there are clearly underlying common design features. For example, mammals, birds, reptiles and fish appear to have similar systems for conserving body nitrogen. All vertebrates and even some insects appear to respond to an increase in protein intake with similar changes in excretory function. Antifreeze protein, a molecule designed to prevent ice crystal damage in body fluids, is present in the antarctic perch and the arctic cod. However, antifreeze protein evolved separately in these two fish. Hence they independently developed a shared solution to a similar environmental challenge. These common design features arise through the process of convergent evolution, which can be simply characterized as similar problems giving rise to similar solutions. However, the best justification for the use of natural animal models is that these models clearly teach us what is biologically possible. As stated by Janine Benyus in her book, "Biomimicry", after 3.8 billion years of evolution nature has learned what works, what is appropriate and what lasts (Benyus, 1997).

This book is organized into six chapters each one dealing with a specific clinical disorder and a description of possible natural animal models for studying that disorder. The choice of clinical disorders has been based simply upon the author's awareness of possible natural animal models for these specific disorders. This book is not meant to be encyclopedic but rather the intent of this book is to foster a comparative physiological approach to clinical problem solving rooted in the use of natural animal models. The clinical disorders and models described in the different chapters serve as examples of this approach.

Each chapter begins with a review of a clinical disorder followed by a discussion of natural animal models for that clinical disorder. The chapters are generally independent of each other and what follows is a brief overview of the clinical disorders covered and a listing of the natural animal models discussed.

The focus of Chapter 1 is on diabetes mellitus, which is becoming almost epidemic in developed countries. Birds, in general, as already mentioned, are a good natural animal model for this disease. Chronic renal failure is reviewed in Chapter 2 and the bear represents a natural model for this disorder. The bear completely recycles its nitrogenous wastes during its dormant period and does not develop any manifestations of renal failure. In Chapter 3 the focus is on atherosclerotic vascular disease. Certain species of fish are a very relevant natural model for this disorder. Salmon develop coronary artery thickenings resembling the early form of mammalian atherosclerosis. Presumably these coronary vascular changes in the salmon are adaptive. However, the lesions in the fish, unlike those in the human, do not show progression. When patients are confined to bed they develop disuse muscle atrophy as well as disuse osteoporosis. These clinical disorders are considered in Chapter 4. The bear is a wonderful natural model for studying these problems since this animal suffers only minor muscle atrophy and osteoporosis during its long dormant period. The problem of ammonia toxicity, which is observed primarily in patients with liver failure or inherited urea cycle enzyme deficiencies, is reviewed in Chapter 5. Fish can be considered a natural animal model since they have evolved adaptations allowing them to tolerate blood ammonia levels that would be lethal in humans. Chapter 6 deals with the problem of hypoxia and ischemia and primarily focuses on the effects of hypoxia/ischemia on the brain. There are a number of natural animal models relevant to this disorder. These models include the turtle, carp and species of birds that fly at high altitudes.

Physiology is a valuable discipline with which to frame important questions concerning the way humans function. Comparative physiology enables us to see that human functions are not unique but are shared at least in part by a variety of animals. The use of natural animal models is simply an extension of this concept. When a human function becomes abnormal because of a disease process, can we find an animal, which has solved a similar problem through the pressures of natural selection? Can that solution be transferred in part or whole to the human with that abnormal function? It is my hope that this book will convince the reader that the answers to these questions are yes.

Reference

Benyus, J.M., 1997. *Biomimicry: Innovation Inspired by Nature.* William Morrow and Company, Inc., New York.

CHAPTER 1

Diabetes Mellitus

Introduction

Diabetes mellitus can be defined as a clinically and genetically hetero-geneous group of disorders characterized by abnormally high levels of glucose in the blood (Harris, 2004). According to the American Diabetes Association Expert Committee (Kahn, 1997), the normal fasting (venous) plasma glucose (FPG) should be less than 6.1 mM (110 mg/dL) where fast-ing is defined as no caloric intake for at least eight hours. A FPG value of 7 mM (126 mg/dL) or more is indicative of diabetes mellitus. Individuals with FPG levels between 6.1 and 7 mM are considered to have impaired fasting glucose. For our purposes, the classification of diabetes mellitus can be simplified to consider two variants only: type 1, which comprises about 5%–10% of cases, and type 2, which comprises about 90% of cases (Harris, 2004). Although the underlying pathogenesis differs between theses two types (insulinopenia in type 1 versus insulin resistance in type 2), the com-plications are similar. These complications include retinal vascular disease, renal glomerulosclerosis, neurological dysfunction and an increased risk of accelerated atherosclerosis. As a consequence of these complications, patients with diabetes have an increased probability of suffering blindness, renal failure, neuropathy, strokes, heart attacks and limb ischemia.

For retinal and renal complications, chronic exposure to high glucose levels is considered the primary causal factor (Chew, 2004; Gruden *et al.*, 2004; Turner, 1998; DCCT research group, 1993). The primary role of chronic hyperglycemia in the pathogenesis of the accelerated atherosclerosis in diabetic patients is still controversial (Semenkovich, 2004; Basta *et al.*, 2004; Brownlee, 2001). In many species of birds, measurement of blood glucose levels have documented values well within the diabetic range using human values as the reference state. Yet these birds do not suffer obvious "diabetic complications." In this chapter, the design features of birds that might explain how this vertebrate tolerates chronic exposure to high blood glucose concentrations without suffering from the severe pathologic consequences observed in humans with diabetes are examined. Understanding the different design features between birds and mammals may give insights into the pathogenesis of diabetic complications or at least allow one to frame new research questions with this purpose in mind.

Blood Glucose Values in Birds

Table 1 summarizes fasting plasma glucose concentrations in various bird species as well as the dietary habits of these birds. This list is not meant to be an exhaustive survey. Beuchat and Chong (1998) have compiled a more complete list of blood glucose levels in birds in the appendix of their paper as well as presenting allometric scaling equations for blood glucose concentrations in both birds and mammals. Many biological and morphological functions scale relative to body size according to allometric equations of the form $y = ax^b$ where y is a biological/morphological variable, a is a proportionality coefficient, x is body mass, and b is the scaling exponent. The allometric equations are given in Table 2.

However, it should be noted that the blood glucose values for birds listed by Beuchat and Chong (1998) in their appendix were not all taken in the fasting state. In some of the referenced studies, the blood glucose levels were measured in birds eating *ad libitum*. However, even feeding should not account for the high blood glucose concentrations commonly observed in birds, since in normal humans even a blood glucose concentration two hours after a meal should not exceed 7.8 mM (Harris, 2004). Nectarivorous birds have extremely high post-feeding blood glucose values, and post-feed levels as high as 42 mM have been measured in Anna's hummingbird (*Calypte anna*) (Beuchat and Chong, 1998).

Table 1 Mean fasting plasma glucose values in birds.

Bird species	FPG (mM)	Dietary habits	Ref.
Great horned owl (*Bubo virginianus*)	20.8	Carnivore	O'Donnell *et al.* (1978)
Red tailed hawk (*Buteo jamaicencis*)	19.3	Carnivore	O'Donnell *et al.* (1978)
Marsh hawk (*Circus cyaneus*)	20.5	Carnivore	O'Donnell *et al.* (1978)
Golden eagle (*Aguila chrysaetos*)	20.5	Carnivore	O'Donnell *et al.* (1978)
Prairie falcon (*Falco mexicanus*)	23.0	Carnivore	O'Donnell *et al.* (1978)
White leghorn chicken (*Gallus domesticus*)	12.1	Carnivore	O'Donnell *et al.* (1978)
Costa's hummingbird (*Calypte costae*)	17.2	Nectarivore	Beuchat and Chong (1998)
Ruby-throated hummingbird (*Archilochus colubris*)	17.8	Nectarivore	Beuchat and Chong (1998)
Anna's hummingbird (*Calypte anna*)	16.7	Nectarivore	Beuchat and Chong (1998)
Palestine sunbird (*Nectarinia osea*)	16.1	Nectarivore	McWhorter *et al.* (2004)
Rook (*Corvus frugilegus*)	13.1*	Omnivore	Miksik and Hodny (1992)

*It is not clear if this is a fasting value. However even a casual value this high in a human would be within the diabetic range.

Several observations can be made from the data in Table 1. First, if we use the American Diabetic Association Expert Committee definition of a normal fasting plasma glucose as less than 6.1 mM, then many bird species have fasting plasma glucose values diagnostic of diabetes mellitus. Secondly, high blood glucose levels are not restricted to nectar-eating birds but cut across all dietary habits. Furthermore, the allometric scaling equations indicate that in general birds have a higher blood glucose concentration than mammals of similar size, and that blood glucose levels are independent of body mass in birds but increase as body mass decreases in mammals. The scaling data reflect the obvious metabolic differences between birds and mammals.

Table 2 Allometric scaling equations.

Biological function (Y)	a	x	b	Reference
Blood glucose concentration mammal (mM)	6.35	kg	−0.104	Beuchat and Chong (1998)
Blood glucose concentration bird (mM)	15.03	kg	−0.014	Beuchat and Chong (1998)
Maximum longevity (yrs) eutherian mammals (captive)	11.6	kg	0.20	Calder (1985)
Maximum longevity (yrs) wild birds	17.6	kg	0.20	Calder (1985)
Maximum longevity (yrs) captive birds	28.3	kg	0.19	Calder (1985)
Resting metabolic rate mammal (kcal/d)	70.5	kg	0.734	Singer (2003)
Resting metabolic rate bird (kcal/day)	86.4	kg	0.668	Singer (2003)
GFR (ml/hr) mammal	1.24	g	0.765	Yokota *et al.* (1985)
GFR (ml/hr) bird	1.24	g	0.694	Yokota *et al.* (1985)
GFR (ml/hr) bird	0.78	g	0.76	Bennett and Hughes (2003)
GFR (ml/hr) bird	0.85	g	0.74	Bakken *et al.* (2004)

Allometric scaling equations have the form $y = ax^b$ where y is the biological function of interest, a is a proportionality coefficient, x is body mass and b is the scaling exponent.

As an additional example underscoring the metabolic differences between birds and mammals, the nectar-feeding bat *Glossophaga soricina* which has a carbohydrate rich diet (although the plant nectar it consumes has a lower sugar concentration than that of hummingbirds (Baker *et al.*, 1998) has a mean fasting plasma glucose value of only 1.4 mM (M. Delorme, personal communication) ($n = 11$). This value is actually quite low since the allometric equation (Table 2) predicts a fasting glucose level of 10.4 mM in this mammal of (mean) body mass 9.1 g.

The higher blood glucose values in birds compared to mammals can be explained by differences in their carbohydrate metabolism. As reviewed by Pollock (2002), the most striking difference between mammals and birds involves the hormonal control of carbohydrate metabolism. Insulin is the dominant pancreatic hormone in mammals whereas in birds glucagon plays that role. In the bird, glucagon circulates at levels ten to 80 times that of mammals (Pollock, 2002). Glucagon is a potent catabolic hormone and stimulates gluconeogenesis and glycogenolysis. The transition to gluconeogenesis is rapid in birds and results from low glycogen stores paired with birds' high metabolic rates. Gluconeogenesis begins several hours postprandially in most birds. Carnivores such as the barn owl exhibit continuous gluconeogenesis from amino acids in both the fed and fasting state (Pollock, 2002). Gluconeogenesis primarily occurs in the liver and more than 70%–75% of all glucose released into the blood stream is made available by the avian liver.

Do Birds Suffer Adverse Consequences from a High Blood Glucose Concentration

Perhaps the strongest evidence that birds do not suffer adverse consequences from higher blood glucose concentrations than mammals relates to their longevity. Beuchat and Chong (1998) comment that hummingbirds can live as long as 12 years in the field. Calder (1985) has examined the comparative biology of longevity using allometric scaling equations for maximum life span in birds and mammals. These equations are given in Table 2. A 3.5 g wild hummingbird has a predicted maximum life span of about six years whereas recorded life spans for the 3.5 g broad-tailed hummingbird (*Selasphorus platycercus*) are actually in the range of eight years (Calder, 1985). The allometric scaling equations for maximum life span predict that a bird will live approximately two and a half times longer than a mammal of similar size. Calder also computed the metabolic life span for birds and mammals, which is defined as the rate of energy turnover multiplied by the maximum life span. Mass specific resting metabolic rate was used as the measure of energy turnover. Metabolic life span was about three times greater in birds than mammals. Although these calculations should be considered very approximate, the existing data suggest that birds not only live longer but are also able to utilize much more basal energy throughout that life span than mammals of similar size. Holmes *et al.* (2001) reviewed the comparative biology of aging in birds and noted that many birds live up to three times longer than mammals of equivalent body mass. They also remarked that the slow aging rates typical of birds compared to mammals were paradoxical given their higher metabolic rates, life time energy expenditures, body temperatures and blood glucose values. Hence, the high blood glucose concentration observed in birds is clearly adaptive and the design characteristics of birds "protect" them from developing the pathologic consequences experienced by humans chronically exposed to similar blood glucose levels.

Anatomic and Physiologic Design Features

Two of the principal organs damaged in patients with diabetes are the eye and kidney. For these two organs, the evidence is quite strong that chronic hyperglycemia is the primary event leading to tissue damage. In addition, there are a reasonable amount of data available on avian eye and kidney design features to compare with mammalian data.

Eye

Retinal nutrient supply

Among vertebrate groups, there is a great variability of retinal blood supply. Non-mammalian animals (including the bird) generally have avascular retinas (Wolburg *et al.*, 1999; Chase, 1982). Mammals are rather unique in that blood vessels penetrate the neuroectodermal tissue from the optic disc and form true intraretinal capillaries. However, there are exceptions even among mammals, with the guinea pig having an avascular retina. Chase (1982) presented data in mammals on the maximum retinal thickness in vascular and avascular retinas. The thickness of avascular retinas does not exceed about 143 microns, which is apparently the maximum oxygen diffusion distance from the choroidal capillaries. Retinas thicker than about 143 microns cannot be nourished by the choroidal capillaries alone and hence require an intraretinal capillary network.

The structure of the avian eye is unusual in view of the situation in mammals. Avian retinas are about 300 microns thick (Chase, 1982; Pettigrew *et al.*, 1990), yet this vertebrate has an avascular retina. Chase's data with respect to mammals would predict that the bird should have developed an intraretinal capillary network in view of the thickness of its retina. However, the bird has developed a unique device called the pecten (Wolburg *et al.*, 1999). The pecten oculi is a convolute of blood vessels projecting from the retina at the exit of the optic nerve into the vitreous chamber toward the lens. In the adult chicken for example, the pecten consists mainly of two cell types: endothelial cells lining blood vessels and pigmented cells intimately associated with the former and filling spaces between blood vessels (Wolburg *et al.*, 1999). The morphological features of the pecten are generally similar in different bird species with some modifications (Smith *et al.*, 1996).

The evidence that the pecten serves a nutritional function for the inner parts of the avascular retina is reviewed by Wolburg *et al.* (1999) and Pettigrew *et al.* (1990). Pettigrew *et al.* (1990) also described in five species of birds, extravasation of fluorescent dye out of pecten vessels across the retinal surface, synchronous with saccadic eye oscillations. This propulsive mechanism would aid in the delivery of nutrients from pecten to retina, which otherwise would occur by diffusion only.

Glucose (and other nutrients) can be delivered to the retinal neurons from the choroidal capillaries via the retinal pigment epithelium. In the majority of mammalian species, the choroidal capillaries alone cannot supply sufficient nutrients for the metabolic needs of the retinal neurons and an intraretinal capillary network develops. The combined choroidal and intraretinal glucose and oxygen supply meets these metabolic needs. In birds, the metabolic demands of the retina are as high as those of mammals, but in this vertebrate the pecten oculi functions to deliver large amounts of glucose and other nutrients via the vitreous. The pecten, then, obviates the necessity for an intraretinal vascular network in the bird.

The functional properties of pecten oculi have been extensively studied in the chicken (Wolburg *et al.*, 1999; Gerhardt *et al.*, 1996) and this data are presented in this section. Presumably, these observations apply to avian pecten in general. In the chicken, the pigmented cells of the pecten are neuroectodermal in origin derived from the retinal pigment epithelium and as such are considered glial cells (Wolburg *et al.*, 1999). These cells contain the enzyme glutamine synthetase. This enzyme serves a very important function in providing glutamine for transmitter synthesis and in removing excess amounts of glutamate and ammonia in the retina. Since there is no metabolic barrier between the vitreous body (with the pecten oculi) and the retina, pectinate endothelial cells must establish a blood retinal barrier in the same way as is done by intraretinal blood vessels in mammals. These endothelial cells express complex tight junctions, and also express the glucose transporter isoform Glut 1 (Gerhardt *et al.*, 1996). In addition, pectinate blood vessels are impermeable to the electron dense tracer lanthanumnitrate. Wolburg *et al.* (1999) suggest that glucose is transported through the pectinate endothelial cells via the Glut 1 transporter into the vitreous and then dispersed over the retinal surface by both diffusion and the mechanical agitation of the pecten during saccadic eye movements, as described by Pettigrew *et al.* (1990). As already noted, mammals (with a few exceptions) are rather unique in having an intraretinal capillary network. In most mammals, an avascular retina is only seen during embryogenesis. During the more advanced developmental stages of the retina of (most) mammals, blood vessels penetrate the neuroectodermal tissue from the optic disc and form true intraretinal capillaries.

Retinal vascularization in mammals

The development of intraretinal capillaries would appear to be a response to the extraordinary metabolic needs of the mammalian neurons. Factors which might be responsible for the development of this intraretinal vascularization are discussed in the paper of Wolburg *et al.* (1999). One of these factors is vascular endothelial growth factor (VEGF), which is secreted by glial cells although other factors are clearly involved. Wolburg *et al.* (1999) point out that although hypoxia has been thought to be the main stimulus to VEGF secretion and subsequent retinal vascularization, there are observations inconsistent with this mechanism. For example, in the guinea pig, the partial pressure of oxygen is extremely low across most of the thickness of the retina (Yu *et al.*, 1996), yet the retina of this mammal remains avascular. Wolburg *et al.* (1999) postulate that in the majority of mammalian species, the choroidal oxygen supply is insufficient to support an oxidative metabolism of the differentiating neurons in the developing retina. The cells are forced to maintain the inefficient glycolytic type of metabolism with a very high rate of glucose consumption. The authors speculate that a glucose deficiency develops, which is the signal stimulating glial cells to secrete VEGF in order to create a vascularization of the retina. As a result, the combined choroidal and intraretinal glucose and oxygen supply becomes sufficient to sustain oxidative (or at least mixed oxidative/glycolytic) energy metabolism of retinal neurons. In birds, the need for intraretinal vascularization is obviated by the pecten oculi. The avian bloodstream has a high blood glucose concentration, and the pecten can pump vast amounts of glucose into the vitreous. This allows the retinal neuronal cells to maintain a predominantly glycolytic mode of energy production preventing the development of a high oxygen demand.

In summary, the avian retina is at least as thick and metabolically active as that of mammals. Unlike mammals, however, birds have evolved a different solution to the problem of nourishing the retina. A special organ, the pecten oculi, supplies nutrients and oxygen to the inner retinal layers — the retina itself remains avascular. A comparison of the physiological features between mammalian and avian retinas might help unravel the primary events lending to diabetic retinopathy.

Avian and mammalian retinal physiology and development

In the following discussion only vascular mammalian retinas will be considered. Under normal physiological conditions, glucose is an essential metabolic substrate of the retina and glycogen stores within the retinal neurons are inadequate to meet basal metabolic demands (Kumagai, 1999). In the mammal, glucose is transported across the endothelial cells of the intraretinal capillaries and from the choroidal vessels across the retinal pigment epithelium. In the bird, glucose is transported across the retinal pigment epithelium from the choroidal vessels but also across the endothelial cells of the pecten oculi. Movement of nutrients out of the pecten towards the retina is aided by the propulsive actions of saccadic eye movements. In mammals, the pathologic changes associated with diabetic retinopathy are localized to the intraretinal microvasculature. Early changes include loss of pericytes, development of acellular non-perfused capillaries, formation of microaneurysms (hypercellular outpouchings) and thickening of the capillary basement membrane (Chew, 2004). In more advanced stages, capillaries become more permeable with leakage of plasma proteins and new vessels form and proliferate (neovascularization). Given that glucose is such a critical metabolite for neural tissue and that inadequate delivery of glucose to retinal glial cells may be the signal initiating mammalian retinal vascularization, how do mammalian and avian retinas compare in their handling of this molecule? In both birds and mammals, the choroidal vessels are leaky and the outer blood-retinal barrier is formed by the retinal pigment epithelial cells (Gerhardt *et al.*, 1996; Kumagai, 1999). These cells are connected by tight junctions and possess glucose transporters (Glut 1 isoform). In the mammal, the inner blood-retinal barrier is formed by the endothelial cells of the intraretinal capillaries while in the bird the pectinate endothelial cells form a blood-retinal barrier. Both the mammalian and avian endothelial cells express the glucose transporter isoform Glut 1.

However, the cellular distribution of this transporter differs between these two vertebrates. In the endothelial cells of the human inner blood-retinal barrier, approximately 50% of total cellular Glut 1 resides in the cytoplasm (Kumagai, 1999) whereas in the pectinate endothelial cells of the chicken, the glucose transporter is restricted to the luminal and abluminal membranes. No transporters were detected in the cytoplasm

(Gerhardt *et al.*, 1996). In mammals, an asymmetrical distribution of Glut 1 has also been documented between the luminal and abluminal endothelial membranes with a ratio of luminal to abluminal transporters of about 1:4 (Farrell and Pardridge, 1991; Kumagai *et al.*, 1996). In the pectinate endothelial cells of the bird, an asymmetrical distribution of Glut 1 between luminal and abluminal membranes is not found (Gerhardt *et al.*, 1996). The different spatial distribution of glucose transporters in the pectinate endothelial cell compared to the human intraretinal endothelial cell is likely a reflection of significant differences in the movement of glucose into and across these two endothelial cells.

Muller cells, a specialized glial cell, which are oriented radially within the retina, also express the Glut 1 isoform both in mammals and birds (Kumagai, 1999; Wolburg *et al.*, 1999). This cell appears to function as a nutrient-supporting cell of the retinal neurons in both mammals and birds (Kumagai, 1999; Wolburg *et al.*, 1999). However there are insufficient data as to whether the nutrient-supporting role is similar in these two vertebrates.

Vascular endothelial growth factor (VEGF) is one of a number of growth factors regulating vascular growth and differentiation. VEGF is a specific mitogen for endothelial cells as well as inducing an increase in endothelial permeability (Schmidt and Flamme, 1998). There is also accumulating evidence that VEGF functions in the nervous system to influence the properties of neural progenitor cells, as well as differentiation and survival of neurons (Hashimoto *et al.*, 2003). VEGF thus participates in multiple distinct biological processes during development. This multifunctionality is an example of the economy of animal design characteristics.

VEGF has been described in a number of animal groups including the human, mouse, rat, cat, quail, and chicken (Schmidt and Flamme, 1998; Stone *et al.*, 1995; Hashimoto *et al.*, 2003). In the human, five different isoforms are known. VEGF binds to specific cell surface tyrosine kinase receptors thus activating kinase activity. VEGF has two known receptors — VEGFR-1 and VEGFR-2 (Flk 1) (Schmidt and Flamme, 1998; Grant *et al.*, 2004). Flk1 is believed to be the key receptor for developmental angiogenesis, hematopoesis, as well as the permeability and neural differentiation effects of VEGF (Grant *et al.*, 2004; Hashimoto *et al.*, 2003).

In the developing avian (chick) retina, ganglion cells appear to be the major site of VEGF synthesis. Later on, after retinal neurogenesis is more complete, VEGF is expressed by both the ganglion cells and by Muller (glial) cells (Hashimoto *et al.*, 2003). The tyrosine kinase receptor Flk1 is expressed initially in developing retinal neuronal cells and following completion of retinal histogenesis Flk1 receptor is detected in Muller glial cells. Since the avian retina remains avascular, VEGF in this tissue is clearly involved in the process of neurogenesis (retinal neuronal differentiation) and not angiogenesis. On the other hand in the developing mammalian retina, VEGF plays roles in both angiogenesis and neurogenesis. In the mouse, at early stages of development VEGF is detected in the ganglion cell layer, and later in the Muller glial cells. The receptor Flk1 is expressed early on in differentiating neural progenitor cells and later on in Muller glial cells (Yang and Cepko, 1996). Thus similar complementary expression patterns of Flk1 and VEGF are present in the developing chick and mouse retinas. Also in both the mature avian (chick) and mammalian (mouse) retinas, Muller cells express both Flk1 and VEGF, thus resulting in an autocrine circuit (Hashimoto *et al.*, 2003; Yang and Cepko, 1996). The significance of this autocrine circuit is unclear.

In the mammal, intraretinal vascularization occurs late in development (Stone *et al.*, 1995; Yang and Cepko, 1996). Astrocytes expressing VEGF spread across the innermost axon layer at approximately the same time as a superficial layer of vessels form and extend across the surface of the retina. Somewhat later a deeper layer of vessels also form and this process is associated with VEGF expression by Muller cells. Stone *et al.* (1995) also demonstrated that the VEGF receptor Flk1 is expressed by endothelial cells in the forming vessels. As discussed by Stone *et al.* (1995) astrocytes only enter retinas in which retinal vasculature will form. Avian retinas lack astrocytes (Schmidt and Flamme, 1998) and as well intraretinal vessels.

As already discussed, the pectinate endothelial cells form a blood-retinal barrier analogous to the blood-retinal barrier formed by the intraretinal vessels of mammals. Pectinate endothelial cells grow by angiogenesis from the ophthalmotemporal artery into the pecten primordium and consecutively gain barrier properties (Wolburg *et al.*, 1999). This vascularization process must be regulated by growth factors and the expression of appropriate receptors on the endothelial cells of the forming blood vessels. However, there appears to be no data as to whether pectinate endothelial cells express Flk1. Schmidt and Flamme (1998) observed that over-expression of (quail) VEGF in the chicken embryo eye caused hypervascularization of vessels of the pecten oculi. Thus it would seem reasonable to assume that the process of vascularization of the pectin oculi is similar to that of vascularization of the mammalian retina with VEGF being a critical growth factor in both birds and mammals.

Thus the avian and mammalian retinas share many similarities. In both, VEGF expressed by ganglion and Muller cells appear to play a crucial role in retinal neurogenesis. In the mammal, VEGF (expressed by invading astrocytes and Muller cells) plays a critical role in intraretinal vascularization. In the bird, the blood vessels of the pecten oculi can be considered the avian equivalent of the mammalian intraretinal blood vessels and the observation of Schmidt and Flamme (1998) suggests that VEGF also regulates vascularization of the pecten.

Finally, several observations support the notion that the microenvironment of the avian retina may actually inhibit vascularization. Firstly, long-term over-expression of (quail) VEGF in the chicken embryo eye failed to cause retinal vascularization (Schmidt and Flamme, 1998). Interestingly, Schmidt and Flamme point out that VEGF stimulates angiogenesis in the cornea (an avascular tissue) of rabbits and small rodents but not in the chicken embryo cornea, which behaves like the chick embryo retina. Secondly, complete intraocular ablation of the pecten oculi in chickens did not result in vascularization of the retina (Brach, 1975).

The design features of the avian and mammalian retinas discussed in this section are summarized in Table 3.

Diabetic retinopathy

In humans, the changes associated with diabetic retinopathy are localized to the intraretinal vessels. Why do similar pathologic changes not occur in the pectinate vessels of the bird, given that a high blood glucose concentration is a design feature of this vertebrate?

As noted in the "Introduction", chronic exposure to high blood glucose levels is considered the primary causal factor in the development of diabetic retinal microvascular disease in humans. Currently, there are four main hypotheses as to how hyperglycemia results in microvascular tissue damage (Brownlee, 2001). These hypotheses reflect which metabolic pathways are believed to be responsible for the tissue damage and include: increased polyol pathway flux, increased advanced glycation end product formation, activation of protein kinase C (PKC) isoforms, and increased hexosamine pathway flux. In all cases, an elevated intracellular glucose concentration or increased intracellular delivery of glucose is considered the factor initiating increased flux through these various biochemical pathways. Specifically with respect to diabetic retinal microvascular disease, the primary event must be an increase in glucose uptake by the retinal microvascular endothelial cells (which comprise the inner blood-retinal barrier). This idea is supported by an observation reported by Kumagai et al. (1996). These investiga-

Table 3 Comparison of avian and mammalian retinas.

Feature	Mammal	Bird
Inner blood-retinal barrier	Intraretinal endothelial cells	Pecten occuli endothelial cells
Outer blood-retinal barrier	Retinal pigment epithelium	Retinal pigment epithelium
Glucose transporters	Intraretinal endothelial cells (50% in cytoplasm; density on abluminal membrane > luminal membrane) Muller cells	Pectinate endothelial cells (none in cytoplasm; density abluminal membrane = luminal membrane) Muller cells
Neurogenesis	Mediated by VEGF* secreted by ganglion and Muller cells	Mediated by VEGF* secreted by ganglion and Muller cells
Angiogenesis	Development of intraretinal vessels mediated by VEGF secreted by astrocytes and Muller cells	Development of pectinate vessels mediated by? VEGF secreted by? Muller cells

*VEGF: Vascular endothelial growth factor.

tors measured glucose transporter densities (using quantitative immuno-gold electron microscopy) in the retinal microvasculature of autopsy specimens from longstanding diabetics. Importantly, none of the eye specimens showed histological evidence of diabetic retinopathy. Compared to controls, the diabetic retinas showed a dramatic increase in the density of Glut 1 transporters in the luminal and abluminal membranes and in the cytoplasm of microvascular endothelial cells. In addition, the asymmetrical distribution of transporters (abluminal membrane > luminal membrane) present in control vessels was lost in the diabetic microvasculature. Such an upregulation of glucose transporters in the microvascular endothelial cells would increase cellular uptake of glucose, thereby delivering more substrate (glucose) to the metabolic pathways responsible for the tissue damage in diabetic patients (Brownlee, 2001). This observation of Kumagai *et al.* (1996) is supported by the study of Knott *et al.* (1996). These authors reported that the expression of glucose transporters (Glut 1 and Glut 3) by cultured normal human retinal endothelial cells was increased in response to exposure to a high glucose concentration (15 mM) in the medium.

By comparison, the endothelial cells of the avian pectinate blood vessels display a spatial distribution of glucose transporters significantly different from that of mammalian intraretinal endothelial cells either from normals or from patients with diabetes. (See "Avian and Mammalian Retinal Physiology and Development" section.) Likely, this difference in transporter localization reflects a difference in glucose movement into and across the pectinate endothelial cell compared to the mammalian intraretinal endothelial cell. More importantly, does this difference mean that in the bird despite a high blood glucose concentration, less glucose is delivered intracellularly to the "pathogenic" metabolic pathways than occurs in the diabetic human? Such an explanation could account for the absence of microvascular tissue damage within the pecten oculi.

The potential for more advanced proliferative damage is also present in the bird since pectinate vessels do respond to over-expression of VEGF by hypervascularization (Schmidt and Flamme, 1998). In humans, proliferative diabetic retinopathy with neovascularization is characterized by upregulation of VEGF expression by retinal cells (Pe'er *et al.*, 1995). Muller cells are probably the cell of origin of the VEGF. Why then does the avian retina not respond to over-expression of VEGF or pecten oculi ablation by vascularization? Avian Muller cells express VEGF and in the mammal Muller cell expression of VEGF correlates with intraretinal vascularization and also with the neovascularization of proliferative diabetic retinopathy in humans.

In summary, much could be learned about the pathogenesis of diabetic retinopathy by studying selected aspects of avian retinal physiology as given in the following examples.

(1) In humans, chronic high blood glucose levels lead to upregulation of glucose transporters in intraretinal endothelial cells resulting in increased cellular uptake and delivery of glucose to the metabolic processes responsible for the vascular damage. No such sequence of events appears to occur in the avian pectinate vessels even though birds have chronically high blood glucose concentrations. Several questions follow from these observations. For example, future research studies could compare the abundance, regulation, and activity of glucose transporters from avian pectinate endothelial cells and from human (normals, diabetic patients) intraretinal endothelial cells exposed to normal and high

glucose concentrations. The absence of "diabetic" changes in pectinate vessels could be either because excess glucose is *not* delivered to the "pathogenic" metabolic processes or these metabolic processes in birds are not activated to the same extent by glucose availability as they are in mammals.

(2) In the developing mammalian retina, Muller cell expression of VEGF is associated with neurogenesis and intraretinal vascularization but only with neurogenesis in the developing avian retina. In the human, pro-liferative diabetic retinopathy with neovascularization is characterized by increased expression of VEGF, probably by Muller cells. In both the mature avian and mammalian retinas, Muller cells express both VEGF and Flk1 forming an autocrine circuit. Given these apparent similarities between avian and mammalian Muller cells, why is Muller cell expression of VEGF in the bird not associated with retinal vascularization? Do the VEGF-Flk1 autocrine circuits in the bird and mammal have similar or dissimilar functions?

Muller cells also express the Glut 1 transporter but there are no data as to whether this transporter is upregulated in diabetes. However, could upregulation of this transporter in diabetes be the initiating step which through increased intracellular delivery of glucose triggers a series of events resulting in increased expression of VEGF by Muller cells with subsequent neovascularization? If so, perhaps in the bird, Muller cell glucose transporters have an activity profile which by limiting glucose uptake prevents the sequence of events leading to increased VEGF expression.

(3) Over-expression of VEGF in the chick embryo eye does not lead to vascularization although VEGF can stimulate angiogenesis in the mammalian cornea, which is also an avascular tissue. Are there properties (biochemical, structural) of the avian retina that inhibit the actions of VEGF?

Kidney

Nephropathy is one of the major complications of diabetes in humans and chronic exposure to high blood glucose concentrations is considered a primary causal factor (Gruden *et al.*, 2004; Turner, 1998; DCCT research group, 1993). Some of the earliest renal abnormalities are an increase in glomerular filtration rate (GFR), renal plasma flow (RPF) and kidney size compared to non-diabetic controls. These changes have been described in type 1

(Christiansen *et al.*, 1982; Parving *et al.*, 2004) and type 2 diabetic patients (Nelson *et al.*, 1996; Vora *et al.*, 1992; Parving *et al.*, 2004) as well as in animal models of diabetes (Thomson *et al.*, 2004a). Renal plasma flow may or may not increase to the same extent as GFR. The pathogenesis of the increased GFR is still unclear (Parving *et al.*, 2004). Total glomerular capillary surface area is increased early in the course of diabetes (Osterby and Gundersen, 1975; Kroustrup *et al.*, 1977; Morgensen *et al.*, 1979) and Morgensen *et al.* (1979) have proposed that the increased filtration surface area might be the main mechanism for the elevated GFR in diabetes. This morphological explanation for the elevated GFR would hold only on the assumption that all other factors affecting GFR (hydrostatic and colloidal osmotic pressure, hydraulic conductivity, renal plasma flow) are unchanged. However, there are data that the increased glomerular capillary surface area cannot fully explain the elevated GFR and hence the other determinants of GFR must also be changed in the diabetic patient.

Christiansen *et al.* (1982) described nine newly diagnosed type 1 diabetic patients who had a mean increase in GFR of 44% compared to healthy controls. After eight days of intensive insulin therapy, with near normal blood glucose control GFR fell but still remained about 20% above the values of the healthy controls. However, the enlargement of the glomerular filtration surface area in type 1 diabetes patients is not reversed by three to seven weeks of insulin therapy (Kroustrup *et al.*, 1977). Hence there is a component (perhaps 50% of the elevated GFR) that is reversed by therapy and is not explicable solely on the basis of the increased glomerular filtration surface area. In addition, Christiansen *et al.* (1981) reported that an intravenous infusion of glucose in normal humans increased GFR and RPF although the changes were modest (6% and 11%, respectively). Volume expansion with saline in the same subjects caused no change in GFR or RPF. Therefore the elevated GFR observed in diabetic patients is the result of an increase in glomerular capillary surface area plus an additional mechanism altering one or more of the other determinants of GFR.

Currently this other mechanism is considered to be a reduced tubuloglomerular feedback signal (Thomsen *et al.*, 2004b). The diabetic kidney displays a primary increase in proximal tubular fluid and electrolyte reabsorption (Vallon, 2003; Thomson *et al.*, 2001) thereby presenting a reduced concentration of Na^+, K^+ and Cl^- to the macula densa cells of the juxtaglomerular apparatus. The resultant decreased uptake of these ions by these cells elicits an increase in the GFR of that nephron via a reduction in vascular tone predominantly of the afferent arteriole.

The GFR data in birds are more difficult to interpret. In humans, the GFR of patients with diabetes can be compared to a matched control group of non-diabetics. In the case of the bird, no such comparable studies are available. In fact, it is not clear what would be an appropriate control group since birds in general have a high blood glucose concentration as compared to mammals. In birds, considering the data available, the only reasonable comparison is that between measured values of GFR and that expected on the basis of body mass alone calculated from allometric scaling equations.

McWhorter *et al.* (2004) measured renal function in a group of Palestine sunbirds (*Nectarinia osea*). These birds (mean body mass 5.8 g) had plasma glucose levels of 16 mM fasting and 28 mM fed. Mean GFR for the group was 1.98 ml/hr. Allometric scaling equations relating GFR to body mass of birds are given in Table 2. For a bird of mass 5.8 g, the expected GFR would be between 2.97 and 4.20 ml/hr depending upon which equation is used. Anna's hummingbird (*Calypte anna*) has a measured GFR of 2.4 ml/hr (Bennett and Hughes, 2003) but an expected GFR based on a body mass of 5.1 g of between 2.69 and 3.84 ml/hr. This bird has average plasma glucose values of 36 mM fed and 16.7 mM fasting (Beuchat and Chong, 1998). A second species of hummingbird, *Selasphorus platycercus* (body mass 3.6 g) has an average GFR of 2.3 ml/hr (Bakken *et al.*, 2004) but an expected GFR of 2.1 to 3.0 ml/hr for a bird of this mass. Blood glucose concentrations have not been measured in this hummingbird but it is reasonable to assume that the values would be similar to those reported for three hummingbird species by Beuchat and Chong (1998) (Table 1 of this chapter). Hence the GFRs measured in these nectarivorous birds are lower than the GFRs expected on the basis of body mass alone. This conclusion is supported by the data of Bakken *et al.* (2004). These authors using a phylogenetically corrected approach found that GFRs for nectarivorous birds fell within the 95% confidence limits for the phylogenetically standardized relationship between GFR and body mass in non-nectarivorous birds.

The inference from all of these observations is that the bird does not suffer from a hyperfiltration state as is found in patients with diabetes. Supporting evidence for this inference comes from a comparison of the avian and mammalian allometric scaling equations for GFR given in Table 2. For equivalent body masses, birds have a lower expected GFR than mammals but a higher blood glucose concentration.

However the use of allometric scaling equations needs to be qualified. These scaling equations are broad paintbrush stroke representations of biological data. The log/log regression line of the equation is the statistical best fit of the data. No single observation will fall exactly on the regression line and it is difficult to know how to interpret the deviations. For example, the herring gull (*Larus argentatus*) has a blood sugar of 19.8 mM. [It is not clear from the original reference of Balasch *et al.* (1974) whether this is a fasting or fed value.] The measured GFR is 264 ml/hr but the expected GFR (based on body mass) is only 148 ml/hr (Bakken *et al.*, 2004).

In summary, at least for the nectar-feeding birds, the evidence appears to be reasonably convincing that the measured GFRs of these birds are not very different from that expected on the basis of body mass alone. Hence the elevated blood glucose concentrations measured in these birds would appear not to lead to an abnormally high GFR as is observed in humans with diabetes.

Kidney anatomy and function in birds and mammals

The renal anatomy and physiology of birds and mammals have been reviewed by Goldstein and Skadhauge (2000), Dantzler (1989), and Kriz and Kaissling (2000). Only selected aspects, relevant to the topic of diabetic nephropathy, will be considered in this section.

Both the bird and mammal show heterogeneity of nephron structure. In the bird, nephrons located in the superficial or cortical region of the kidney are very simple and are often referred to as reptilian-type nephrons (Dantzler, 1989). These nephrons contain a glomerulus, proximal and distal tubules and a collecting duct. They lack loops of Henle. Deeper in the avian kidney lie the mammalian types nephrons. These nephrons have larger, more complex glomeruli and also have a loop of Henle. However, the transition from the reptilian-type nephrons to the mammalian types is gradual, not abrupt, so that there are nephrons of intermediate size and relatively short loops of Henle in the area between the superficial and deeper parts of the kidney (Dantzler, 1989). Of the total nephron population, between 10% and 30% possess loops of Henle (i.e. are of the mammalian type) (Goldstein and Skadhauge, 2000). However, the kidney of Anna's hummingbird (*Calypte anna*) seems to be an outlier in that only 0.4% of nephrons have a loop of Henle (i.e. 99.6% of nephrons are of the reptilian type) (Casotti *et al.*, 1998). Whether this extremely low complement of looped nephrons is a unique feature of this particular species or a general characteristic of nectarivorous birds is unknown. The mammalian kidney displays less nephron heterogeneity. All nephrons contain the same elements (glomerulus, proximal convoluted and straight tubuli, loop of Henle, distal convoluted tubule, connecting tubule, and collecting duct) but differ according to the length of the loop of Henle (Dantzler, 1989; Kriz and Kaissling, 2000).

Each glomerulus in the avian or mammalian kidney is supported by an afferent arteriole, which breaks up into a glomerular capillary network. The complexity of the glomerular capillary network is much greater in mammals than birds (Dantzler, 1989; Beuchat *et al.*, 1999). Unlike the mammal, the avian kidney has a second afferent blood supply, the renal portal system. The blood that enters the kidney via the renal portal system mixes with post-glomerular efferent arteriolar blood in peritubular sinuses and eventually flows to the renal veins. There is no evidence that renal portal flow contributes directly to glomerular filtration under physiological conditions (Dantzler, 1989). In most non-mammalian vertebrates (including birds), changes in whole-kidney GFR apparently result primarily from changes in the number of glomeruli filtering (glomerular intermittency). Vertebrates with intermittent glomerular filtration have a renal portal system that can continue to nourish the cells of non-filtering nephrons in the absence of a post-glomerular arterial supply (Dantzler, 1989). By contrast in mammals changes in whole kidney GFR involve changes in the filtration rates of all nephrons (Dantzler, 1989).

The juxtaglomerular apparatus (JGA) is well developed in the mammal (Schnermann and Briggs, 2000; Vallon, 2003). The end of the thick ascending limb of Henle's loop runs through an angle generated by the afferent and efferent glomerular arterioles and thus makes contact with the vascular pole of its own glomerulus. The JGA comprises all structures that form this site of contact and represents a functional and structural link between: (1) the macula densa cells, specialized tubular cells at the end of the thick ascending limb of Henle's loop; (2) cells of the extraglomerular mesangium (Goormaghtigh or lacis cells), which fill the angle between the afferent and efferent glomular arteriole (the macula densa cells abut upon this cushion of cells); (3) the vascular smooth muscle cells and renin-secreting cells (juxtaglomerular granular cells) located in the media of the afferent arteriolar wall. The JGA apparatus regulates the GFR of its own nephron through the mechanism of tubuloglomerular feedback. This mechanism refers to the series of events whereby changes in the ion (Na^+, K^+, Cl^-) concentrations in the tubular fluid are sensed by the macula densa via the Na^+-K^+-$2Cl^-$ co-transporter in its luminal membrane. An increase or decrease in ion uptake elicits inverse changes in GFR by altering vascular tone, chiefly of the afferent arteriole. Even though most avian nephrons lack a loop of Henle, there is good evidence that the bird kidney has a complete JGA (Dantzler, 1989). Morild *et al.* (1985) studied five domestic fowl and noted all components of the JGA in both reptilian and mammalian nephrons at all cortical levels. The components identified included macula densa cells and extraglomerular mesangial cells in the angle between the afferent and efferent arterioles. The only incomplete feature was that granular epithelioid cells were only found occasionally in the afferent arterioles of mammalian type JGAs. They did not observe granular epithelioid cells in the afferent arteriole of reptilian-type glomeruli or in any efferent arterioles of either type of nephron. However, in other studies by Christensen *et al.* (1982) and Sokabe and Ogawa (1974) the full JGA has been noted to be present: juxta glomerular granular cells, macula densa cells, extraglomerular mesangial cells, and attachment of the distal tubule to the vascular pole of the glomerulus. Hence a JGA is not a unique feature of the mammalian kidney. In fact, distal nephrons are closely associated with their corresponding glomerular vascular pole in amphibians and reptiles as well (Dantzler, 1989) suggesting that tubular regulation of single nephron GFR may be a general feature of most vertebrates.

Diabetic nephropathy

The role of changes in the expression or activity of glucose transporters in the pathogenesis of diabetic intraretinal vascular disease has already been reviewed. There is also accumulating evidence that changes in the expression or activity of renal glucose transporters may be the initiating step in the sequence of events resulting in diabetic kidney disease.

(a) Renal glucose transporters

The bulk of filtered glucose is reabsorbed in the proximal convoluted tubule by low affinity high capacity glucose transporters SGLT2 and Glut 2 (Marks *et al.*, 2003). SGLT2 is a sodium-glucose co-transporter, which carries Na^+ down its electrochemical gradient and glucose against its concentration gradient across the brush border membrane. Accumulated glucose then exits the cell across the basolateral membrane via a Na^+-independent mechanism involving the facilitated glucose transporter Glut 2. SGLT1 and Glut 1 are high affinity low capacity transporters found in the straight portion of the proximal tubule and are responsible for the reabsorption of the remaining glucose presented to this later portion of the proximal tubule. Marks *et al.* (2003) examined glucose transporter activity in brush border membrane (BBM) vesicles in a rat diabetic model compared to a non-diabetic control. They observed that SGLT mediated glucose transport was unchanged by diabetes. However, they found that diabetes increased facilitative glucose transport at the BBM by 67.5% and that this effect was accompanied by significantly raised levels of Glut 2 protein at the BBM. Normally Glut 2 is not found in the BBM. They also observed that Glut 2 protein was undetectable in the BBM when the diabetic animal fasted overnight with restoration of blood glucose levels to normal. This finding implies that a raised level of glucose in the plasma or tubular fluid is the stimulus for the expression of Glut 2 in the BBM. Vestri *et al.* (2001), also using a diabetic animal model, found that the diabetic state compared to the control state was associated with an increased abundance of SGLT1, SGLT2, GLUT2 mRNAs as well as an increase in GLUT2 protein. Interestingly, Glut 1 mRNA and protein abundance was decreased in diabetic animals compared to controls.

Presumably, an elevated plasma or tubular fluid glucose concentration leads to a change in the expression (usually an increase) of proximal tubule glucose transporters. These changes probably represent responses designed to adjust the transtubular glucose flux. However, these changes in glucose transporter expression appear to be maladaptive and result in increased delivery of glucose to metabolic processes involved in the genesis of tissue damage characteristic of diabetic renal disease (Marks *et al.*, 2003). The sequence of events is thus similar to that responsible for diabetic retinopathy.

There are no studies in the bird, which have explicitly looked for renal glucose transporters. Kono *et al.* (2005) did report expression of mRNA for Glut 1, Glut 2, Glut 3 and Glut 8 in the chicken kidney with Glut 2 and Glut 8 mRNA having the highest expression. Sweazea and Braun (2006) also described Glut 1 and Glut 3 gene expression in the kidney of English sparrows (*Passer domesticus*). However, the data of McWhorter *et al.* (2004) on renal function in Palestine sunbirds (*Nectarinia osea*) give a measure of the capacity of the avian kidney for glucose reabsorption. In these birds, mean GFR was 1.98 ml/hr and mean fed glucose concentration was 28.18 mM. Hence glucose filtered load (FL) was 0.056 mmoles/hr. In the fed state, glucose concentration in ureteral urine averaged 2.97 mM and urine flow rate was about 0.36 ml/hr. Hence, the sunbird reabsorbed about 98% of filtered glucose and the glucose reabsorption rate was 0.055 mmoles/hr. In comparison, the normal human has a maximum glucose reabsorption capacity (Tm) of about 125 mmoles/hr (Pitts, 1968). The mean body mass of the Palestine sunbirds was 5.8 g and we will assume the human weighs 70,000 g. How do we compare the glucose reabsorption rates of these two vertebrates? As noted in the "Blood Glucose Values in Birds" section, many biological functions scale relative to body size according to an allometric equation of form $y = ax^b$. One of these biological functions is resting metabolic rate for which the scaling exponent b is 0.75 (Schmidt-Nielsen, 1984). It is clearly inaccurate to compare the avian and human glucose reabsorption rates by dividing each by its respective body mass, since this method presupposes a linear relationship between this biological function and body mass. In comparative physiology, biological functions are often normalized by dividing the value of the function by metabolic body mass ($x^{0.75}$). This is the preferred method. The Palestine sunbird's metabolic body mass is 3.74 g and the human's metabolic body mass is 4304 g. The normalized glucose reabsorption rates are 0.015 mmoles/g $^{0.75}$/hr for the sunbird and 0.029 mmoles/g $^{0.75}$/hr for the human. Hence on a (metabolic) body mass basis, the human has a renal tubular glucose reabsorption rate about double that of the Palestine sunbird. The reabsorptive capacity in a human with diabetes is probably enhanced due to increased expression of glucose transporters, so that the difference between the glucose reabsorption rates of the diabetic human and the Palestine sunbird would be even greater.

The mechanism of glucose transport in the avian kidney has not been directly studied but most likely is similar to that of other vertebrates (Dantzler, 1989); sodium mediated tubular cell uptake (luminal side) and facilitated diffusion out of the cell (peritubular side). Assuming that metabolic body mass is an appropriate quantity for normalizing avian and human renal function, then the transcellular glucose flux in the Palestine sunbird kidney would be substantially less than that of the diabetic human. The implication would be that in the sunbird renal tubules, less glucose would be delivered to intracellular metabolic processes which have been incriminated in the genesis of tissue damage, than in the diabetic human.

Mammalian glomerular mesangial cells also express glucose transporters, the predominant one being Glut 1 (Heilig *et al.*, 1997a). Cultured rat mesangial cells exposed to 20 mM glucose compared to 8 mM displayed a significant increase in Glut 1 mRNA and Glut 1 protein. These changes were associated with enhanced glucose uptake by these cells (Heilig *et al.*, 1997b). In addition, cultured rat mesangial cells exposed to a high glucose concentration also had higher rates of synthesis of extracellular matrix (ECM) (Heilig *et al.*, 1997a; Heilig *et al.*, 1995). Interestingly this enhanced synthesis of extracellular matrix (ECM) was also observed in cultured mesangial cells over-expressing the glucose transporter Glut 1 but exposed to a normal (8 mM) glucose concentration (Heilig *et al.*, 1995). These results clearly indicate that increased glucose transport activity with an increase in delivery of intracellular glucose is the critical step for enhanced ECM synthesis. Enhanced mesangial cell ECM synthesis would clearly play a critical role in the genesis of diabetic glomerulosclerosis. The role played by a high extracellular glucose concentration is to increase glucose transport activity by inducing an increased expression of the Glut 1 transporter.

(b) Juxtaglomerular apparatus

The elevated GFR observed in the early stages of diabetes has already been discussed. This hyperfiltration is accompanied by kidney growth, which correlates temporally with increased ornithine decarboxylase (ODC) activity (Kaysen *et al.*, 1989; Thomson *et al.*, 2001). ODC is the rate-controlling enzyme for the biosynthesis of growth regulatory polyamines. The larger kidney in diabetic patients has a larger glomerular capillary filtering surface area as well as a larger tubular mass. The greater tubular mass and increased expression of sodium-glucose co-transporters found in the diabetic are responsible for increased proximal tubular fluid and electrolyte reabsorption. A reduced concentration of ions is presented to the macula densa cells suppressing the glomerulotubular feedback signal and increasing GFR. Avian kidneys contain a complete juxtaglomerular apparatus and are capable of autoregulating GFR when renal artery pressure is changed (Wideman and Gregg, 1988). The presence of glomerulotubular feedback has not been directly examined in birds (Dantzler, 1989). However, Vena *et al.* (1990) have presented data that tubuloglomerular feedback may not contribute significantly to avian GFR autoregulation. The sensitivity or gain of the tubuloglomerular feedback mechanism can be reduced in mammals by adaptation to increased salt intake and/or volume expansion (Schnermann and Briggs, 2000; Vena *et al.*, 1990). Vena *et al.* (1990) observed that GFR autoregulatory profiles were identical in birds fed a high or low sodium diet. However, these results may indicate that the tubuloglomerular feedback mechanism in birds has very different settings and adaptation properties than the tubuloglomerular feedback mechanism of mammals.

However, given that birds possess a complete JGA and that a tubuloglomerular feedback mechanism has been documented in amphibia (Persson and Persson, 1981; Persson *et al.*, 1989), it would appear quite reasonable to assume a tubuloglomerular feedback mechanism also exists in birds. Why then do birds not have a hyperfiltration state as present in humans with diabetes? In fact why do birds not develop "diabetic" renal damage? Birds have chronically high blood sugar values, which in the human increase the expression of glucose transporters. Unfortunately, there are insufficient data available on avian renal physiology to answer these questions. Studies need to address at least the following questions.

(1) Are there glucose transporters in the avian kidney and more specifically in glomerular mesangial and proximal tubular cells?

The answer is clearly yes given that mRNA for four Glut isoforms are expressed in the chicken and sparrow kidney (Kono *et al.*, 2005; Sweazea and Braun, 2006) and that glucose transporters have been detected in a variety of other avian tissues including endothelial cells of the pecten (Glut 1; Gerhardt *et al.*, 1996), erythroblasts (Glut 1, Glut 3; Johnstone *et al.*, 1998), liver (Glut 2; Wang *et al.*, 1994), and small bowel (SGLT 1; Dyer *et al.*, 1997b and Barfull *et al.*, 2002).

However the intrarenal location, abundance and activity profile of these kidney transporters are unknown. Once avian renal glucose transporters have been defined, their properties should be compared to those of mammals exposed to both normal and high glucose concentrations?

(2) Is a glomerulotubular feedback mechanism present in the avian kidney?

If so, does this glomerulotubular feedback mechanism have different "settings" from that of mammals? Is the rate of proximal tubular glucose reabsorption in the bird low enough after feeding (see in "Renal Glucose Transporters section" "Diabetic Nephropathy") that proximal electrolyte reabsorption is not increased above that of the fasting state and suppression of the glomerulotubular feedback signal does not occur in contrast to the situation in the diabetic mammal?

Glycated hemoglobin

Glucose can condense spontaneously and non-enzymatically with proteins. These glycated proteins subsequently degrade into ketoaldehydes that can cross-link to form advanced glycation end products. Because the condensation reaction is non-enzymatic, the rate of protein glycation is strongly influenced by the ambient time-averaged glucose concentration (Beuchat and Chong, 1998). Hence, glycated hemoglobin (HbA$_{1C}$) has been used as an integrated measure of chronic blood glucose levels in diabetic patients (Genuth, 2004). Glycated hemoglobin levels have been measured in birds and have been found to be low (compared to mammals) despite the generally high blood glucose values observed in birds (Beuchat and Chong, 1998; Rendell *et al.*, 1985). However, interpretation of this observation is not straightforward. Higgins *et al.* (1982) measured glycated hemoglobin in a variety of mammals and found that the levels were determined by three major variables: plasma glucose concentration, red blood cell lifespan, and red blood cell glucose permeability. For example, the pig has a blood glucose concentration similar to that of a normal human but an almost undetectable glycated hemoglobin level. This difference is due to the very low permeability of pig erythrocytes to glucose (Higgins *et al.*, 1982).

In fact, among mammals there is considerable variation in erythrocyte glucose permeability and this variation reflects differences in the expression of glucose transporters. Human red cells show a high density of glucose transporters (chiefly Glut 1) whereas pig erythrocytes have virtually no glucose transport activity (Johnstone *et al.*, 1998). In the pig, erythrocytes lose glucose transporters during the maturation process. In birds, there are data that red blood cells have a shorter lifespan than those of mammals (Beuchat and Chong, 1998) and that avian erythrocytes also have a lower glucose permeability than human erythrocytes (Shields *et al.*, 1964). This reduced permeability can be accounted for by a loss of glucose transporters at an early stage of red blood cell maturation. In the chicken, loss of glucose transporters (Glut 1 and Glut 3 isoforms) occurs prior to the appearance of the cells in the peripheral circulation (Johnstone *et al.*, 1998). Mature avian erythrocytes display a low level of glucose transport activity (Johnstone *et al.*, 1998), which is probably accounted for by the few remaining transporters. However, there is evidence that glucose transport in the erythrocytes of the adult bird may also occur via a calcium dependent mechanism (Bihler *et al.*, 1982).

Although glycated hemoglobin levels are low in birds, plasma levels of glycated albumin are not. Rendell *et al.* (1985) found glycated albumin levels in the chicken, duck and turkey to be much higher than the levels measured in a variety of mammals.

One further piece of the puzzle is that chronic hyperglycemia in humans (i.e. diabetes) increases the density of glucose transporters in erythrocyte membranes (Harik *et al.*, 1991). Hence in the human, hyperglycemia induces an increase in the expression of glucose transporters in the eye, kidney, and on erythrocytes. An important, but currently unanswerable question is whether the low glucose permeability of avian erythrocytes compared to the human is a feature of this cell type alone or is a low glucose permeability a property of avian cells in general?

Interestingly, Barfull *et al.* (2002) found that the density of the sodium-glucose co-transporter SGLT1 in chicken jejunum brush border membrane was 40% less in the adult than in the two-day-old chicken. This decrease was not the result of a change in diet or intraluminal glucose concentration. This is an important observation since the sugar content of the diet can modify the activity and abundance of SGLT1 intestinal transporters in mammals (Barfull *et al.*, 2002; Dyer *et al.*, 1997a). In addition, diabetes is associated with an increase in abundance of intestinal glucose transporters in humans (Dyer *et al.*, 2002) and in streptozotocin induced diabetic rats (Miyamoto *et al.*, 1991). Therefore for chicken enterocytes, transcellular glucose flux would be less in the adult than in the newborn. This finding is similar to the situation in avian erythrocytes although erythrocytes lose transporters during maturation whereas enterocytes lose transporters later after hatching. Hence in avian erythrocytes and enterocytes, glucose transporters are decreased in abundance in the mature compared to immature animal. By contrast in the diabetic mammal, there is an increased expression of erythrocyte and enterocyte glucose transporters compared to non-diabetic mammals.

Concluding Remarks and Future Research

In the mammal and in particular the human, a coherent picture is beginning to emerge with respect to at least the eye and kidney complications of diabetes. There are four main hypotheses as to how chronic hyperglycemia causes tissue damage.

These hypotheses reflect which metabolic pathways are believed to be responsible for the tissue damage and include: increased polyol pathway flux, increased advanced glycation end product formation, activation of protein kinase C isoforms and increased hexosamine pathway flux. Advanced glycation end products can have direct effects on extracellular matrix and lipids as well as receptor-mediated biological effects (Basta *et al.*, 2004). Recently Brownlee (2001) has proposed a unifying explanation, pointing out that each of these four different pathogenic mechanisms reflects a single hyperglycemia induced process — over-production of superoxide by the mitochondrial electron-transport chain.

However, the primary initiating event linking hyperglycemia to the metabolic pathways culminating in tissue damage is the delivery of an increased amount of glucose to these metabolic processes. This primary event involves an increased expression of glucose transporters in the eye and kidney. Hyperglycemia induces such an increased expression by an unknown mechanism in intraretinal vascular endothelial cells as well as in glomerular mesangial and renal proximal tubular cells. Glucose cellular uptakes are thereby enhanced and more glucose is made available to the aforementioned metabolic pathways. Hyperglycemia is also associated with increased growth of the kidney but the role of glucose transporters in this process is not known. The metabolic basis for kidney growth is thought to involve increased synthesis of growth regulatory polyamines.

Kidney growth results in an increase in glomerular capillary surface area and in tubular mass. The elevated GFR, an early feature of diabetes, is due to both this increased filtering surface area plus a reduced glomerulo-tubular feedback signal as a result of increased proximal tubular fluid and electrolyte reabsorption. The increased GFR is believed to be a contributory factor to progressive renal damage (Thomson *et al.*, 2004b).

The bird represents a wonderful natural model for this disease. Birds in general have higher blood glucose levels than mammals. Nectarivorous birds, for example, have fasting values of about 20 mM and fed values as high as 42 mm. These blood glucose concentrations are similar to those inducing increased expression of glucose transporters in mammals. How do birds avoid the problem of diabetic complications? This natural model tells us that a solution to this problem exists. Birds show us that it is biologically possible to have chronically high blood glucose levels without developing pathological consequences so prevalent in humans with diabetes. The initiating step in the pathogenesis of complications in mammals appears to be an increased expression of glucose transporters induced by hyperglycemia. This observation raises the question as to whether diabetes can be considered, at least in part, a disorder of glucose transporters. It would also seem logical to examine if modification of this step is the critical adaptation in the bird. Simplistically, modification of glucose transporters (density and/or activity) would be a much more economical adaptation than modification of the many metabolic pathways leading to tissue damage. Unfortunately the data available to examine this question are very sparse.

Birds have the same glucose transport isoform (Glut 1) on pecten oculi endothelial cells as mammals have on the endothelial cells of intraretinal vessels. The cellular spatial distribution of this transporter differs between mammals and birds. However, the density and activity profile of the avian transporter have not been examined. In the bird, does this transporter function to limit the influx of glucose into the cell and hence limit glucose availability to the various metabolic pathways? In the human, the glucose transporter (Glut 1) on erythrocytes is more abundant in the diabetic patient than in the normal human and the expression of intestinal glucose transporters is also increased in diabetic mammals compared to non-diabetics. However in the bird, erythrocyte and intestinal glucose transporters have a decreased density in the mature compared to the immature animal. Is a low level of glucose transport activity unique to avian erythrocytes and enterocytes or is it a property of avian glucose transporters in general?

In mammals, hyperglycemia increases expression of renal glucose transporters (mesangial cells, proximal tubular cells). In the bird, glucose transporter mRNA is expressed in the kidney but the intrarenal location and properties of these transporters have not been described. Both avian and mammalian kidneys have a complete juxtaglomerular apparatus and a glomerulotubular feedback mechanism well documented in the mammal, is most likely operative in the bird as well.

Increased proximal reabsorption of ions and glucose (via sodium-glucose co-transporters) in the diabetic mammal reduces the concentration of ions delivered to the macula densa, thereby suppressing the glomerulotubular feedback signal and increasing GFR. The nectar-eating Palestine sunbird appears to have a lower post-feeding renal tubular glucose reabsorptive rate than the normal human, but it is also plausible that the avian glomerulotubular feedback mechanism has different settings than that of the normal mammal.

In summary, the design features of birds that allow this vertebrate to have a chronically high blood glucose concentration without the occurrence of pathological consequences are currently unknown. Existing data give us tantalizing clues but are very incomplete. The need for more research is obvious and the following is a list of possible questions. Several of these questions were considered in more detail previously in the "Diabetic Retinopathy" section.

(1) How does the density and activity profile of the glucose transporter on the endothelial cells of the pecten oculi compare to that of the glucose transporter on the endothelial cells of the mammalian intraretinal vessels? This study needs to be done under conditions in which the

mammalian endothelial cell transporters are exposed to both normal and high extracellular glucose concentrations.

(2) As discussed in the section on the eye, Muller cells probably play a role in the neovascularization stage of diabetic retinopathy. Since Muller cells in both birds and mammals secrete vascular endothelial growth factor (VEGF) and express the VEGF receptor Flk1, why does neovascularization not occur in the bird? Are Muller cell glucose transporters in the mammal upregulated in diabetes and if so, is this event linked to increased VEGF expression with subsequent neovascularization? If so, are the properties of avian Muller cell glucose transporters such that this sequence of events does not occur in the bird?

Does the autocrine circuit (VEGF-Flk1 receptor) have a different function in the avian compared to the mammalian Muller cell? Are there biochemical and/or structural properties of the avian retina that inhibit the actions of VEGF?

(3) Glucose transporters in the avian kidney need to be characterized with respect to intrarenal location, abundance and activity profile. The properties of these avian transporters need to be compared to those of the mammal under conditions in which the mammalian transporters are exposed to both normal and high extracellular glucose concentrations.

(4) Does the bird have a glomerulotubular feedback mechanism? Given the observation that the bird has a complete juxtaglomerular apparatus, the answer to this question is most likely yes. If yes, studies should be conducted to define the properties or settings of this mechanism. It would be important to compare the settings of the avian glomerulotubular feedback mechanism with that of the normal and the diabetic mammal.

The specific design features by which the bird avoids the pathologic consequences of a chronically high blood glucose concentration are currently a mystery. By solving this mystery and thus characterizing these design features, we could then examine the feasibility of inducing such design features in humans with diabetes.

Given the very limited data currently available, a reasonable working hypothesis is that the properties (density and activity profile) of avian glucose transporters result in a lower rate of cellular glucose uptake then observed in mammals with diabetes. Hence in the bird compared to the diabetic mammal, less glucose would be delivered to various "pathogenic" metabolic pathways and tissue damage would be avoided. The observation

that avian adult erythrocytes and adult enterocytes have less glucose transport activity than erythroblasts and newborn enterocytes is consistent with
this working hypothesis. By contrast, glucose transport activity in erythrocytes and enterocytes is increased in the diabetic mammal compared to the
non-diabetic. In addition, the following observations are also supportive
of this tentative hypothesis.

Compared to mammals, birds are considered to be insulin resistant (Fan
et al., 1993; Tokushima *et al.*, 2005). Although the mechanism responsible
for this resistance has not been determined, one possibility is that this
resistance reflects differences in the effects of insulin on cellular glucose
uptake in mammals and birds. In mammals, the glucose transporter Glut 4
is the insulin responsive isoform that is found primarily in tissues (skeletal
muscle, heart, adipocytes) that respond to elevations in circulating insulin
levels with a rapid increase in glucose uptake and metabolism (Mueckler,
2004). In the basal state, Glut 4 is primarily localized to intracellular storage
sites. Binding of insulin to its receptor in the plasma membrane initiates
a cascade of events resulting in the translocation of Glut 4 transporters
to the plasma membrane with a concomitant reduction in the amount in
the intracellular pool (Hou *et al.*, 2004). Chicken skeletal muscle *in vivo*
responds to exogenous insulin with an increase in glucose uptake, although
the magnitude of this effect is less than occurs in rodents (Tokushima *et al.*,
2005). Interestingly chickens lack the Glut 4 transporter (Tokushima *et al.*,
2005) so the mechanism underlying insulin stimulated glucose transport
in chicken skeletal muscle is unknown. Skeletal muscle from the English
sparrow (*Passer domesticus*) also lacks the Glut 4 transporter (Sweazea and
Braun, 2006). However, the Glut 4 transporter has been detected in the
gastrocnemius muscle of Muscovy ducklings (Thomas-Delloye *et al.*, 1999).
In this bird, a supra-maximal concentration of exogenous insulin increased
glucose uptake *in vitro* in the perfused lower limb by five fold. In comparison, a perfused rat hindlimb preparation demonstrated a 13 fold increase in
glucose uptake using the same experimental protocol. This difference in the
effectiveness of insulin between duckling and rat underscores the insulin
resistance of avian skeletal muscle. Sarcolemmal vesicles isolated from
insulin-stimulated duckling gastrocnemius muscle showed an increase in
glucose uptake as well as an increase in amount of Glut 4 transporters
compared to vesicles isolated from unstimulated muscle. The increased
amount of Glut 4 in insulin-stimulated vesicles was due to translocation
of transporters from intracellular sites to the sarcolemma. Hence, insulin
appears to have similar qualitative effects on glucose uptake in avian and

mammalian skeletal muscle but quite different quantitative effects. The greater effect of insulin in stimulating glucose uptake in mammalian skeletal muscle compared to skeletal muscle of the bird must reflect presently undefined differences in the properties of mammalian and avian Glut 4 transporters.

As a rough approximation, this working hypothesis can be summarized as follows. In the mammal, blood glucose concentrations are tightly regulated and only in the diabetic state with upregulation of glucose transporters does the intracellular concentration of glucose become excessive. In the bird, blood glucose concentrations are higher than in the mammal and probably less tightly regulated. The bird controls the intracellular concentration of glucose by limiting glucose influx via glucose transporters. However, currently there are no published data as to whether and if so how avian and mammalian glucose transporters differ. Hopefully, future experimental studies in birds will actually test this working hypothesis which in light of present day knowledge must be considered still very speculative.

References

Baker, H.G., Baker, I., Hodges, S.A., 1998. Sugar composition of nectars and fruits consumed by birds and bats in the tropics and subtropics. *Biotropica* 30, 559–586.

Bakken, B.H., McWhorter, T.J., Tsahar, E., Martinez del Rio, C., 2004. Hummingbirds arrest their kidneys at night: diel variation in glomerular filtration rate in *Selasphorus platycerus*. *J. Exp. Biol.* 207, 4383–4391.

Balasch, J., Palomeque, J., Palacios, L., Musquera, S., Jiminez, M., 1974. Hematological values of some great flying and aquatic-diving birds. *Comp. Biochem. Physiol.* 49A, 137–145.

Barfull, A., Garriga, C., Mitjans, M., Planas, J.M., 2002. Ontogenetic expression and regulation of Na^+-D-glucose cotransporter in jejunum of domestic chicken. *Am. J. Physiol. Gastrointest. Liver Physiol.* 282, G559–G564.

Basta, G., Schmidt, A.M., DeCaterina, R., 2004. Advanced glycation end products and vascular inflammation: implications for accelerated atherosclerosis in diabetes. *Cardiovascular Res.* 63, 582–592.

Bennett, D.C., Hughes, M.R., 2003. Comparison of renal and salt gland function in three species of wild ducks. *J. Exp. Biol.* 206, 3273–3284.

Beuchat, C.A., Chong, C.R., 1998. Hyperglycemia in hummingbirds and its consequences for hemoglobin glycation. *Comp. Biochem. Physiol. Part A,* 120, 409–416.

Beuchat, C.A., Preest, M.R., Braun, E.J., 1999. Glomerular and medullary architecture in the kidney of Anna's hummingbird. *J. Morphol.* 240, 95–100.

Bihler, I., Charles, P., Sawh, P.C., 1982. Sugar transport regulation in avian red blood cells: role of Ca^{2+} in the stimulatory effects of anoxia, adrenaline, and ascorbic acid. *Can. J. Physiol. Pharmacol.* 60, 615–621.

Brach, V., 1975. The effect of intraocular ablation of the pecten oculi of the chicken. *Invest. Ophthalmol.* 14, 166–168.

Brownlee, M., 2001. Biochemistry and molecular cell biology of diabetic complications. *Nature* 414, 813–820.

Calder III, W.A., 1985. The comparative biology of longevity and lifetime energetics. *Exp. Gerontol.* 20, 161–170.

Casotti, G., Beuchat, C.A., Braun, E.J., 1998. Morphology of the kidney in a nectarivorous bird, the Anna's hummingbird *Calypte anna. J. Zool. Lond.* 244, 175–184.

Chase, J., 1982. The evolution of retinal vascularization in mammals. A comparison of vascular and avascular retinae. *Ophthalmology* 89, 1518–1525.

Chew, E.Y., 2004, Pathophysiology of diabetic retinopathy. In: LeRoith, D., Taylor, S.I., Olefsky, J.M. (Eds.), *Diabetes Mellitus: A Fundamental and Clinical Text,* 3rd Edn. Lippincott Williams and Wilkins, Philadelphia, PA, pp. 1303–1314.

Christiansen, J.S., Frandsen, M., Parving, H-H., 1981. Effect of intravenous glucose infusion on renal function in normal man and in insulin-dependent diabetics. *Diabetologia* 21, 368–373.

Christiansen, J.S., Gammelgaard, J., Tronier, B., Svendsen, P.A., Parving, H-H., 1982. Kidney function and size in diabetics before and during initial insulin treatment. *Kidney Int.* 21, 683–688.

Christensen, J.A., Morild, I., Mikeler, E., Bohle, A., 1982. Juxtaglomerular apparatus in the domestic fowl (*Gallus domesticus*). *Kidney Int.* 22, Suppl. 12, S24–S29.

Dantzler, W.H., 1989. *Comparative Physiology of the Vertebrate Kidney.* Springer-Verlag, Heidelberg, Germany.

The Diabetes Control and Complications Trial research group, 1993. The effect of intensive treatment of diabetes on the development and progression of long-term complications in insulin-dependent diabetes mellitus. *N. Engl. J. Med.* 329, 977–986.

Dyer, J., Barker, P.J., Shirazi-Beechey, S.P., 1997a. Nutrient regulation of the intestinal Na^+/glucose co-transporter (SGLT1) gene expression. *Biochem. Biophys. Res. Comm.* 230, 624–629.

Dyer, J., Ritzhaupt, A., Wood, I.S., de la Horra, C., Illundain, A.A., Shirazi-Beechey, S.P., 1997b. Expression of the Na^+/glucose co-transporter (SGLT1) along the length of the avian intestine. *Biochem. Soc. Trans.* 25, 480S.

Dyer, J., Wood, I.S., Palejwala, A., Ellis, A., Shirazi-Beechey, S.P., 2002. Expression of monosaccharide transporters in intestine of diabetic humans. *Am. J. Physiol. Gastrointest. Liver Physiol.* 282, G241–G248.

Fan, L., Gardner, P., Chan, S.J., Steiner, D.F., 1993. Cloning and analysis of the gene encoding hummingbird proinsulin. *Gen. Comp. Endocrinol.* 91, 25–30.

Farrell, C., Pardridge, W.M., 1991. Blood-brain barrier glucose transporter is asymmetrically distributed on brain capillary endothelial lumenal and ablumenal membranes: an electron microscopic immunogold study. *Proc. Natl. Acad. Sci. USA* 88, 5779–5783.

Genuth, S., 2004. Application of the diabetes control and complications trial. In: LeRoith, D., Taylor, S.I., Olefsky, J.M. (Eds.), *Diabetes Mellitus: A Fundamental and Clinical Text*, 3rd Edn. Lippincott Williams and Wilkins, Philadelphia, PA, pp. 645–657.

Gerhardt, H., Liebner, S., Wolburg, H., 1996. The pecten oculi of the chicken as a new *in vivo* model of the blood-brain barrier. *Cell Tissue Res.* 285, 91–100.

Goldstein, D.L., Skadhauge, E., 2000. Renal and extrarenal regulation of body fluid composition. In: Whittow, G.C. (Ed.), *Sturkie's Avian Physiology*, 5th Edn. Academic Press, San Diego California, pp. 265–297.

Grant, M.B., Afzal, A., Spoerri, P., Pan, H., Shaw, L.C., Mames, R.N., 2004. The role of growth factors in the pathogenesis of diabetic retinopathy. *Expert Opin. Investig. Drugs* 13, 1275–1293.

Gruden, G., Gnudi, L., Viberti, G., 2004. Pathogenesis of diabetic nephropathy. In: LeRoith, D., Taylor, S.I., Olefsky, J.M. (Eds.), *Diabetes Mellitus:*

A Fundamental and Clinical Text, 3rd Edn. Lippincott Williams and Wilkins, Philadelphia, PA, pp. 1315–1330.

Harik, S.I., Behmand, R.A., Arafah, B.M., 1991. Chronic hyperglycema increases the density of glucose transporters in human erythrocyte membranes. *J. Clin. Endocrinol. Metab.* 72, 814–818.

Harris, M.I., 2004. Definition and classification of diabetes mellitus and the criteria for diagnosis. In: LeRoith, D., Taylor, S.I., Olefsky, J.M. (Eds.), *Diabetes Mellitus: A Fundamental and Clinical Text*, 3rd Edn. Lippincott Williams and Wilkins, Philadelphia, PA, pp. 457–467.

Hashimoto, T., Zhang, X., Yang, X., 2003. Expression of the Flk 1 receptor and its ligand VEGF in the developing chick central nervous system. *Gene Expr. Patterns* 3, 109–113.

Heilig, C.W., Concepcion, L.A., Riser, B.L., Freytag, S.O., Zhu, M., Cotes, P., 1995. Over-expression of glucose transporters in rat mesangial cells cultured in a normal glucose milieu mimics the diabetic phenotype. *J. Clin. Invest.* 96, 1802–1814.

Heilig, C.W., Brosius III, F.C., Henry, D.N., 1997a. Glucose transporters of the glomerulus and the implications for diabetic nephropathy. *Kidney Int.* 52, Suppl. 60, S91–S99.

Heilig, C.W., Liu, Y., England, R.L., Freytag, S.O., Gilbert, J.D., Heilig, K.O., Zhu, M., Concepcion, L.A., Brosius III, F.C., 1997b. D-glucose stimulates mesangial cell Glut 1 expression and basal and IGF-1 sensitive glucose uptake in rat mesangial cells. Implications for diabetic nephropathy. *Diabetes* 46, 1030–1039.

Higgins, P.J., Garlink, R.L., Bunn, H.F., 1982. Glycosylated hemoglobin in human and animal red cells. Role of glucose permeability. *Diabetes* 31, 743–748.

Holmes, D.J., Fluckiger, R., Austad, S.N., 2001. Comparative biology of aging in birds: an update. *Exper. Gerontol.* 36, 869–883.

Hou, J.C., Saltiel, A.R., Pessin, J.E. 2004. Cellular and molecular processes regulating glucose transporter 4 translocation. In: Le Roith, D., Taylor, S.I., Olefsky, J.M. (Eds.), *Diabetes Mellitus: A Fundamental and Clinical Text*, 3rd Edn. Lippincott Williams and Wilkins, Philadelphia, PA, pp. 349–364.

Johnstone, R.M., Mathew, A., Setchenska, M.S., Grdisa, M., White, M.K., 1998. Loss of glucose transport in developing avian red cells. *Eur. J. Cell Biol.* 75, 66–77.

Kahn, R., 1997. Report of the expert committee on the diagnosis and classification of diabetes mellitus. *Diabetes Care* 20, 1183–1197.

Kaysen, G.A., Rosenthal, C., Hutchinson, F.N., 1989. GFR increases before renal mass or ODC activity increase in rats fed high protein diets. *Kidney Int.* 36, 441–446.

Knott, R.M., Robertson, M., Muckersie, E., Forrester, J.V., 1996. Regulation of glucose transporters (Glut 1 and Glut 3) in human retinal endothelial cells. *Biochem. J.* 318, 313–317.

Kono, T., Nishida, M., Nishiki, Y., Seki, Y., Sato, K., Akiba, Y., 2005. Characterization of glucose transporter (Glut) gene expression in broiler chickens. *Br. Poult. Sci.* 46, 510–515.

Kriz, W., Kaissling, B., 2000. Structural organization of the mammalian kidney. In: Seldin, D.W., Giebisch, G. (Eds.), *The Kidney, Physiology and Pathophysiology*, 3rd Edn. Lippincott Williams and Wilkins, Philadelphia, PA, pp. 587–654.

Kroustrup, J.P., Gundersen, H.J.G., Osterby, R., 1977. Glomerular size and structure in diabetes. III. Early enlargement of the capillary surface. *Diabetologia* 13, 207–210.

Kumagai, A.K., Vinores, S.A., Pardridge, W.M., 1996. Pathological upregulation of inner blood-retinal barrier Glut 1 glucose transporter expression in diabetes mellitus. *Brain Res.* 706, 313–317.

Kumagai, A.K., 1999. Glucose transport in brain and retina: implications in the management and complications of diabetes. *Diabetes Metab. Res. Rev.* 15, 261–273.

Marks, J., Carvou, N.J., Debnam, E.S., Srai, S.K., Unwin, R.J., 2003. Diabetes increases facilitative glucose uptake and Glut 2 expression at the rat proximal tubule brush border membrane. *J. Physiol.* 553(1) 137–145.

McWhorter, T.J., Martinez del Rio, C., Pinshow, B., Roxburgh, L., 2004. Renal function in Palestine sunbirds: elimination of excess water does not constrain energy intake. *J. Exp. Biol.* 207, 3391–3398.

Miksik, I., Hodny, Z., 1992. Glycated hemoglobin in mute swan (*Cygnus olor*) and Rook (*Corvus frugilegus*). *Comp. Biochem. Physiol.* 103B, 553–555.

Miyamoto, K., Hase, K., Taketani, Y., Minami, H., Oka, T., Nakabou, Y., Hagihira, H., 1991. Diabetes and glucose transporter gene expression in rat small intestine. *Biochem. Biophys. Res. Comm.* 181, 1110–1117.

Morgensen, C.E., Osterby, R., Gundersen, H.J.G., 1979. Early functional and morphologic vascular renal consequences of the diabetic state. *Diabetologia* 17, 71–76.

Morild, I., Mowinckel, R., Bohle, A., Christensen, J.A., 1985. The juxtaglomerular apparatus in the avian kidney. *Cell Tissue Res.* 240, 209–214.

Mueckler, M., 2004. Gemline manipulation of glucose transport and glucose homeostasis. In: Le Roith, D., Taylor, S.I., Olefsky, J.M. (Eds.), *Diabetes Mellitus: A Fundamental and Clinical Text*, 3rd Edn. Lippincott Williams and Wilkins, Philadelphia, PA, pp. 431–439.

Nelson, R.G., Bennett, P.H., Beck, G.J., Tan, M., Knowler, W.C., Mitch, W.E., Hirschman, G.H., Myers, B.D., 1996. Development and progression of renal disease in Pima Indians with non-insulin dependent diabetes mellitus. *N. Engl. J. Med.* 335, 1636–1642.

O'Donnell, J.A., Garbett, R., Morzenti, A., 1978. Normal fasting plasma glucose levels in some birds of prey. *J. Wildl. Dis.* 14, 479–481.

Osterby, R., Gundersen, H.J.G., 1975. Glomerular size and structure in diabetes mellitus. 1. Early abnormalities. *Diabetologia* 11, 225–229.

Parving, H.-H., Mauer, M., Ritz, E., 2004. Diabetic nephropathy. In: Brenner, B.M. (Ed.), *Brenner and Rector's the Kidney*, 7th Edn. Saunders, Philadelphia, PA, pp. 1777–1818.

Pe'er, J., Shweiki, D., Itin, A., Hemo, I., Gnessin, H., Keshet, E., 1995. Hypoxia-induced expression of vascular endothelial growth factor by retinal cells is a common factor in neovascularizing ocular diseases. *Lab. Invest.* 72, 638–645.

Persson, B.-E., Persson, A.E.G., 1981. The existence of a tubulo-glomerular feedback mechanism in the Amphiuma nephron. *Pflugers Arch.* 391, 129–134.

Persson, B.-E., Sakai, T., Ekblom, M., Marsh, D.J., 1989. Effect of bumetanide on tubuloglomerular feedback in *Necturus maculosus*. *Acta Physiol. Scand.* 137, 93–99.

Pettigrew, J.D., Wallman, J., Wildsoet, C.F., 1990. Saccadic oscillations facilitate ocular perfusion from the avian pecten. *Nature* 343, 362–363.

Pitts, R.F., 1968. *Physiology of the Kidney and Body Fluids*, 2nd Edn. Year Book Medical Publishers, Chicago, Illinois.

Pollock, C., 2002. Carbohydrate regulation in avian species. *Semin. Avian Exotic Pet Med.* 11, 57–64.

Rendell, M., Stephen, P.M., Paulsen, R., Valentine, J.L., Rasbold, K., Hestorff, T., Eastberg, S., Shint, D.C., 1985. An interspecies comparison of normal levels of glycosylated hemoglobin and glycosylated albumin. *Comp. Biochem. Physiol.* 81B, 819–822.

Schmidt, M., Flamme, I., 1998. The *in vivo* activity of vascular endothelial growth factor isoforms in the avian embryo. *Growth Factors* 15, 183–197.

Schmidt-Nielsen, K., 1984. *Scaling, Why is Animal Size So Important?* Cambridge University Press, Cambridge, UK.

Schnermann, J., Briggs, J.P., 2000. Function of the juxtaglomerular apparatus: control of glomerular hemodynamics and renin secretion. In: Seldin, D.W., Giebisch, G. (Eds.), *The Kidney, Physiology and Pathophysiology*, 3rd Edn. Lippincott Williams and Wilkins. Philadelphia, PA, pp. 945–980.

Semenkovich, C.F., 2004. Diabetes mellitus, lipids, and atherosclerosis. In: LeRoith, D., Taylor, S.I., Olefsky, J.M. (Eds.), *Diabetes Mellitus: A Fundamental and Clinical Text*, 3rd Edn. Lippincott Williams and Wilkins, Philadelphia, PA, pp. 1365–1375.

Shields, C.E., Herman, Y.F., Herman, R.H., 1964. I-^{14}C-glucose utilization of intact nucleated red blood cells of selected species. *Nature* 263, 935–936.

Singer, M.A., 2003. Dietary protein-induced changes in excretory function: a general animal design feature. *Comp. Biochem. Physiol. Part B* 136, 785–801.

Smith, B.J., Smith, S.A., Braekevelt, C.R., 1996. Fine structure of the pecten oculi of the barred owl (*Strix varia*). *Histol. Histopathol.* 11, 89–96.

Sokabe, H., Ogawa, M., 1974. Comparative studies of the juxtaglomerular apparatus. In: Bourne, G.H., Danielli, J.F., Jeon, K.W. (Eds.), *International*

Review of Cytology, Volume 37. Academic Press, San Francisco, California, pp. 271–327.

Stone, J., Itin, A., Alon, T., Pe'er, J., Gnessin, H., Chan-Ling, T., Keshet, E., 1995. Development of retinal vasculature is mediated by hypoxia-induced vascular endothelial growth factor (VEGF) expression by neuroglia. *J. Neurosci.* 15, 4738–4747.

Sweazea, K.L., Braun, E.J., 2006. Glucose transporter expression in English sparrows (*Passer domesticus*). *Comp. Biochem. Physiol. Part B* 144, 263–270.

Thomas-Delloye, V., Marmonier, F., Duchamp, C., Pichon-Georges, B., Lachuer, J., Barre, H., Crouzoulon, G., 1999. Biochemical and functional evidences for Glut 4 homologous protein in avian skeletal muscle. *Am. J. Physiol.* 277, R1733–R1740.

Thomson, S.C., Deng, A., Bao, D., Satriano, J., Blantz, R.C., Vallon, V., 2001. Ornithine decarboxylase, kidney size, and the tubular hypothesis of glomerular hyperfiltration in experimental diabetes. *J. Clin. Invest.* 107, 217–224.

Thomson, S.C., Deng, A., Komine, N., Hammes, J.S., Blantz, R.C., Gabbai, F.B., 2004a. Early diabetes as a model for testing the regulation of juxtaglomerular NOS 1. *Am. J. Physiol.* 287, F732–F738.

Thomson, S.C., Vallon, V., Blantz, R.C., 2004b. Kidney function in early diabetes: the tubular hypothesis of glomerular filtration. *Am. J. Physiol. Renal Physiol.* 286, F8–F15.

Tokushima, Y., Takahashi, K., Sato, K., Akiba, Y., 2005. Glucose uptake *in vivo* in skeletal muscles of insulin injected chicks. *Comp. Biochem. Physiol. Part B* 141, 43–48.

Turner, R., 1998. Intensive blood-glucose control with sulphonylureas or insulin compared with conventional treatment and risk of complications in patients with type 2 diabetes (UKPDS 33). *Lancet* 352, 837–853.

Vallon, V., 2003. Tubuloglomerular feedback and the control of glomerular filtration rate. *News Physiol. Sci.* 18, 169–174.

Vena, V.E., Lac, T.H., Wideman Jr., R.F., 1990. Dietary sodium, glomerular filtration rate autoregulation, and glomerular size distribution profiles in domestic fowl (*Gallus gallus*). *J. Comp. Physiol. B.* 160, 7–16.

Vestri, S., Okamoto, M.M., de Freitas, H.S., Aparecida dos Santos, R., Nunes, M.T., Morimatsu, M., 2001. Changes in sodium or glucose filtration rate modulate gene expression of glucose transporters in renal proximal tubular cells of rat. *J. Membr. Biol.* 182, 105–112.

Vora, J.P., Dolben, J., Dean, J.D., Thomas, D., Williams, J.D., Owens, D.R., Peters, J.R., 1992. Renal hemodynamics in newly presenting non-insulin dependent diabetes mellitus. *Kidney Int.* 41, 829–835.

Wang, M.-Y., Tsai, M.-Y., Wang, C., 1994. Identification of chicken liver glucose transporter. *Arch. Biochem. Biophys.* 310, 172–179.

Wideman Jr., R.F., Gregg, C.M., 1988. Model for evaluating avian renal hemodynamics and glomerular filtration rate autoregulation. *Am. J. Physiol.* 254, R925–R932.

Wolburg, H., Liebner, S., Reichenbach, A., Gerhardt, H., 1999. The pecten oculi of the chicken: a model system for vascular differentiation and barrier maturation. *Int. Rev. Cytol.* 187, 111–159.

Yang, X., Cepko, C.L., 1996. Flk-1, a receptor for vascular endothelial growth factor (VEGF), is expressed by retinal progenitor cells. *J. Neurosci.* 16, 6089–6099.

Yokota, S.D., Benyajati, S., Dantzler, W.H., 1985. Comparative aspects of glomerular filtration in vertebrates. *Renal Physiol.* 8, 193–221.

Yu, D., Cringle, S.J., Alder, V.A., Su, E., Yu, P.K., 1996. Intraretinal oxygen distribution and choroidal regulation in the avascular retina of guinea pigs. *Am. J. Physiol.* 270, H965–H973.

Chronic Renal Failure

Introduction

Among mammals, the kidney is of sole importance in regulating the excretion of nitrogenous end products and in regulating the solute and water composition of the internal environment. In other vertebrate groups, extrarenal routes for regulation of solute and water movements or postrenal modification of ureteral urine or both are also important (Dantzler, 1989). For example, in reptiles and birds, renal function is coordinated with the postrenal transport of ions and water across the bladder, cloaca or colon. In some species, extrarenal salt glands play an important role in fluid and electrolyte balance (Dantzler, 1989). The term renal failure usually refers to impairment of global renal function rather than to specific defects for example in the renal tubular transport of certain ions or water. The term chronic, as used in clinical medicine, implies a process of relatively long duration and one that is not readily reversible. In humans, chronic renal failure (CRF) is generally due to a pathologic process resulting in a loss of nephrons. The natural animal model considered in this chapter is the American black bear (*Ursus americanus*). During winter sleep, the bear's physiological situation can be characterized as analogous to chronic renal failure in the human. Before discussing the bear, some core metabolic/physiological features in mammals relevant to an understanding of the problem of chronic renal failure and important as a background for appreciating the bear's particular adaptations are reviewed.

Although the causes of CRF in humans are many, in North America about 60%–70% of new patients requiring dialysis suffer from either diabetes or hypertensive vascular disease (Schulman and Himmelfarb, 2004). In this chapter, then, CRF in humans will be considered the functional consequences of irreversible nephron loss. The underlying pathological process resulting in nephron loss may or may not be progressive. In addition, in some patients clinical signs and symptoms may be partly due to the manifestations of renal failure *per se* and partly due to the manifestations of the underlying disease process. In this chapter, discussion will be limited to the manifestations of renal failure only.

The physiology of the failing kidney can be understood in terms of the intact nephron hypothesis (Bricker, 1969). In essence, as renal failure advances, whole kidney function reflects the activities of a diminishing pool of intact nephrons rather than relatively constant numbers of nephrons each with diminishing function. As nephrons are irreversibly lost, the remaining intact nephrons undergo structural and functional adaptations (Taal *et al.*, 2004) which include: an increase in single nephron glomerular filtration rate, nephron hypertrophy (glomerular enlargement and an increase in tubular mass), and an increase in proximal tubular reabsorption such that glomerulotubular balance is maintained. The assembly of signs and symptoms resulting from renal failure are collectively known as the uremic syndrome (Bailey and Mitch, 2004). The basic abnormality is the presence of waste products that are not being eliminated by the kidney. Furthermore, the ultimate source of most so called uremic toxins is dietary protein. Consequently excessive dietary protein will induce uremia in patients with only modestly reduced renal function and uremic symptoms can be relieved when dietary protein is restricted.

For example, Kopple and Coburn (1973) studied the effects of low protein diets (20 and 40 g protein/day) in eight males with severe chronic renal failure; creatinine clearance between 2.5 and 10.5 ml/min/1.73 m^2. Mean nitrogen balance was negative with the 20 g protein diet and neutral or positive with the 40 g diet. Uremic symptomatology (16 different symptoms) improved while the patients were ingesting either the 20 or 40 g protein diet. In a more prolonged study, Nelson *et al.* (1985) followed 15 patients (nine males, six females) with chronic renal failure (creatinine clearance about 10 ml/min/1.73 m^2) prescribed a diet containing 0.38 g protein/kg/day. The protein was of high biological value. Fourteen patients continued in the study for one year, 12 patients for two years, six for three years and three for up to 54 months. Average blood urea concentration remained stable at approximately 16 mM for up to 48 months. Mean hematocrit remained stable and mean serum albumin increased from 2.91 to 3.57 g/dℓ. Lean body mass remained stable and patients reported an improved sense of well-being while on the low protein diet. It should be noted that patients in both of these studies (Kopple and Coburn, 1973; Nelson *et al.*, 1985) suffered from primary renal failure without any evidence of systemic diseases such as diabetes.

Perhaps, the best evidence for the existence of so called uremic toxins is the fact that dialysis is life saving in the acutely uremic patient and that patients with severe chronic renal failure receiving regular dialysis can be maintained in reasonable health for many years.

There are, in addition, specific functions that are compromised in the failing kidney. For example, the synthesis of erythropoietin is reduced and this functional defect is the chief cause of anemia in chronic renal failure. However, the predominant factor in the genesis of the uremic syndrome is the mismatch between the diminished excretory capacity of the kidney and the generation of protein metabolic end products requiring excretion. This concept of a mismatch for the failing kidney underscores the observation that there is normally a close linkage between metabolic demand and renal excretory function. This linkage is discussed in the next section.

Metabolic Rate and Renal Function

Table 1 lists the allometric scaling equations for resting metabolic rate, and excretory functions in mammals, birds, reptiles and fish. Allometric scaling analysis deals with the effects of changes in body size (or scale) on biological functions and structures. Many biological and morphological functions scale, relative to body size, according to allometric equations of the form $y = ax^b$, where y is a biological variable, a is a proportionality coefficient, x is body mass, and b is the scaling exponent. Furthermore, we would expect biological functions that are closely linked to scale to body mass in a similar fashion, i.e. to have similar scaling exponents. Considering that the data in Table 1 encompass a large variety of animals of different size, the similarity in the scaling exponents for resting metabolic rate (RMR) and renal/gill excretory functions both within and across vertebrate groups is very striking.

Table 1   Allometric scaling equations.

Biological function (*y*)	*a*	*x*	*b*
Mammal			
Resting metabolic rate (kcal/d)	70.5	kg	0.734
Whole body protein turnover (g/d)	16.6	kg	0.74
Urinary N excretion (g/d)	0.146	kg	0.72
Urea production (gN/d)	0.52	kg	0.76
rRNA turnover (mg/d)	347	kg	0.69
tRNA turnover (mg/d)	38	kg	0.78
GFR (ml/h)	1.24	g	0.765
Bird			
Resting metabolic rate (kcal/d)	86.4	kg	0.668
GFR (ml/h)	1.24	g	0.694
Reptile			
Resting metabolic rate (O_2 consumption, ml/h, temp. 30°C)	0.278	g	0.77
GFR (ml/h)	0.0058	g	0.999
Fish			
Resting metabolic rate (joules/h, temp. 20°C)	2.5	g	0.70
GFR (ml/h)			
Fresh water	0.045	g	0.647
Salt water	0.0021	g	0.888
Gill surface area (cm^2)	–	g	0.7–0.8

GFR: Glomerular filtration rate. N: nitrogen. The data in this table are from Singer (2002) and Singer (2003a).

Allometric scaling equations have the form $y = ax^b$ where y is the biological function of interest, a is a proportionality coefficient, x is body mass and b is the scaling exponent.

From an evolutionary perspective, this linkage between RMR and renal/gill excretory functions is a highly conserved vertebrate design feature. Since mammals have been the most studied animal group, we can use this vertebrate to further analyze the coupling between RMR and in this case glomerular filtration rate (GFR). In mammals, whole body protein turnover, daily urea production and daily RNA turnover all scale to body size with an exponent very close to that of RMR and GFR (Singer, 2002; Table 1). In addition, in mammals it is estimated that whole body protein turnover accounts for about 20% to 30% of RMR (Darveau *et al.*, 2002; Rolfe and Brown, 1997; Waterlow, 1984). In the mammal (at least), the scaling data describe a linked set of metabolic/physiological processes comprising energy flux (RMR), protein and RNA turnover, rate of urea production, urinary nitrogen excretion rate and GFR (Singer, 2002).

There is also circumstantial evidence in mammals that within this linked set of processes, whole body protein turnover (a significant component of RMR) regulates GFR and hence the rate of nitrogenous waste product excretion (Singer, 2001a and 2003a). Given the allometric scaling data presented in Table 1, it is highly probable that in birds, reptiles and fish as well whole body protein turnover regulates nitrogen waste excretion via regulation of GFR (birds, reptiles, fish), gill excretory rate (fish) and renal tubular secretion of urate (birds) (Singer, 2003a).

In summary, there is reasonably good evidence for the existence of metabolic regulation of renal (or gill) nitrogenous waste excretion. Based on mammalian data, it appears that whole body protein turnover is probably the metabolic component involved in this regulatory process. The proximate signals linking whole body protein turnover and excretory functions are unknown but ammonia may be one of the signals (Singer, 2001b and 2003a). It is also worth stressing that the renal (gill) functions responsible for nitrogenous waste excretion are integrated with other renal (gill) functions such as those responsible for volume and electrolyte balance.

The importance of dietary protein in the genesis of the uremic syndrome and the apparent regulation of renal nitrogenous waste excretion by whole body protein turnover underscore the central role of nitrogen metabolism in the setting of chronic renal failure. The next sections contain a discussion of relevant aspects of nitrogen and urea metabolism and protein turnover in mammals.

Nitrogen and Urea Metabolism and Protein Turnover

Nitrogen and urea metabolism

Nitrogen (N) balance refers to the difference between total N intake and total N losses. Nitrogen intake is generally in the form of dietary protein. There is some variation in N output in the feces, probably depending on the fiber content of the diet and in N loss from the skin, depending on climate. However the chief route of N excretion is via the kidney and in practical terms the N is lost as urea. For example when the protein intake of Holstein heifers was increased, over 90% of the additional N excreted in the urine was in the form of urea (Marini and Van Amburgh, 2003). As pointed out by Waterlow (1999), the balances between N input and output and protein synthesis and degradation are connected through the free amino acid pool as illustrated in Fig. 1. For most amino acids, the catabolic steps resulting

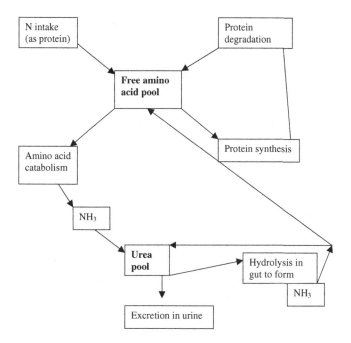

Fig. 1 N and protein balance.

in the net conversion of the alpha amino group to ammonia requires the concerted action of glutamate transaminase and glutamate dehydrogenase (Rodwell, 2000b). The carbon skeleton is further processed, and because the majority of amino acids are gluconeogenic, the end result is conversion to glucose. Ammonia is a toxic molecule (Cooper and Plum, 1987) and in mammals ammonia is detoxified at a considerable energetic cost through the synthesis of urea. A portion of the urea appearing in the urea pool is excreted in the urine and a portion is transferred to the intestine where it is hydrolyzed by bacterial urease. A more detailed review of urea metabolism is presented in a later section.

According to Waterlow (1999) the regulation of urea production is the key to N balance. Urea is synthesized almost exclusively in the liver although a complete set of urea cycle enzymes has been described in pig enterocytes (Davis and Wu, 1998; Wu, 1995). Current evidence suggests that the concentration of ammonia at the site of carbamoyl-phosphate synthetase 1 (the first enzyme in the cycle) is the short-term regulator of urea production (Waterlow, 1999). Long-term regulation of urea production is mediated by changes in enzyme amounts and hence changes in enzyme activity. With increasing dietary protein, the activity (amount) of all five enzymes increases and to very much the same extent. The fact that four of the five genes involved are on separate chromosomes makes the coordination of these changes in enzyme abundance all the more remarkable. Jackson (1998) reported that in humans urea production remains relatively constant over a wide range of N intakes. Waterlow (1999) has recalculated Jackson's data and showed that there was a positive statistical relationship, but not a very sensitive one, between protein intake and urea production. However, for N intakes between 50 and 200 mg N/kg/day (0.3 and 1.25 g protein/kg/day), urea N production exceeds N intake. This observation indicates that over this range of protein intakes (0.3 to 1.25 g/kg/day) if part of the urea produced was not salvaged by the intestinal tract instead of being excreted, it would be impossible to achieve N balance.

As already noted, a portion of produced urea is transferred to the intestine where it is hydrolyzed by bacterial urease, thus generating ammonia as a potential nitrogen source. This topic has been reviewed by Singer (2003b). The fraction of urea produced that is transferred to the intestine and hydrolyzed by bacterial urease in humans is about 20% (Walser and Bodenlos, 1959). What is unclear and still controversial is the fate of ammonia released by intestinal hydrolysis of urea. Simplistically, one can consider several paths for this ammonia. A portion can be used by the intestinal microflora as an N source for the synthesis of amino acids and proteins. A portion could be absorbed into the portal venous system and returned to the liver. There the ammonia could either enter the ornithine cycle to reform urea or be used in the synthesis of non-essential amino acids. In the ruminant, transfer of urea into the gastrointestinal tract with subsequent recovery of urea N is well documented to be very important for maintenance of N balance during periods of low N intake (Singer, 2003b). Table 2 summarizes urea kinetics in Holstein heifers fed isocaloric diets of varied N content (Marini and Van Amburgh, 2003). In cows ingesting a low N diet (1.45%), 83% of produced urea was recycled to the intestine, whereas only 29% was recycled when the cows had a high N intake (3.40%). The fraction of recycled urea N used for anabolism remained fairly constant across diets, from 59.8% to 60% for heifers on the 1.45% and 3.40% N diets, respectively. About 43% of recycled urea N was used for microbial protein synthesis by ruminal bacteria in animals ingesting the 1.45% N diet, whereas the value was only 5% when the animals received a high N diet (3.40% N). Urea pool size decreased and turnover of the urea pool was faster as nitrogen intake decreased.

Table 2 Urea kinetics in Holstein heifers.

	1.45% N diet	3.40% N diet
N intake (g/d)	87.6	203.5
UNP* (g/d)	31.1	135.2
UNI† (g/d)	25.9	39.5
UNI returned to ornithine cycle (g/d)	5.0	13.1
N derived from UNI incorporated into microbial proteins by ruminal bacteria‡	11.0	2.0
UNI used to synthesize non-essential amino acids by host (g/d)	4.5	21.7
UNI excreted in feces (g/d)	5.3	2.7

*UNP urea N production.
†UNI urea N transferred to gastrointestinal tract.
‡In this study UNI is a measure of urea N recycled to the total gastrointestinal tract. Urea N incorporated into microbial protein was measured for ruminal bacteria only.

In a comparable study done in lambs, there was a linear decrease in ruminal, duodenal, ileal and cecal attached bacterial urease activity with increasing N intake (Marini *et al.*, 2004).

Hence, the ruminant possesses a very sophisticated system. Urea is transferred to the gastrointestinal tract and hydrolyzed with release of ammonia. Since amino acids are absorbed chiefly in the small intestine (Mayes, 2000), important synthetic steps occur upstream in the rumen. Ruminal bacteria are capable of synthesizing amino acids and protein from simple sources of N such as ammonia. These microbial derived proteins can then be absorbed as amino acids through the small intestine after digestion (Stevens and Hume, 1998). This system is most important at times of limited N intake and its versatility was demonstrated by Virtanen (1966). Ruminal bacteria are capable of synthesizing all the amino acids (including essential amino acids) necessary for protein synthesis and cows fed a diet of purified carbohydrates with urea and ammonium salts, as the sole sources of N were well maintained and capable of relatively high milk production. The regulatory mechanisms coordinating the various steps in this system are presently unknown.

Is a similar system operative in non-ruminant mammals? Although there is agreement that urea is transferred to the gastrointestinal tract in non-ruminants and hydrolyzed by bacterial urease there is disagreement over the fate of the ammonia so liberated. Furthermore, although colonic bacteria, like ruminal microbes, can synthesize amino acids and proteins using ammonia as an N source (Singer, 2003b) since the small bowel is the major site of amino acid absorption, amino acids derived from microbial synthesis in the cecum/colon may not be readily available to the non-ruminant host. What then is the evidence that urea cycling to the gastrointestinal tract represents an important N salvaging system in non-ruminant mammals?

Jackson (1998) has stated that in humans most of the N derived from urea hydrolysis is retained within the metabolic pool of nitrogen with only 10%–20% returning directly to urea formation. However, Young and El-Khoury (1994) have come to the opposite conclusion based upon their own studies in humans. They believe that the major fate of the ammonia that is released via urea hydrolysis is its return to the urea cycle, effectively resulting in a "futile" cycle although they do not rule out a physiological purpose of such a process under certain circumstances. However, a review of published data in humans and other non-ruminant mammals supports the notion that not all the ammonia generated via urea hydrolysis is recycled back to urea, and that a significant fraction of this released ammonia is used for synthetic purposes particularly under conditions of a limited N intake (Singer, 2003b). In fact, recent studies to be described show that amino acids synthesized by the intestinal microflora of non-ruminants can be absorbed by the host.

Unlike bacteria, mammals cannot incorporate inorganic nitrogen into lysine (an essential amino acid), which does not engage in transamination. Torrallardora *et al.* (1996) confirmed this by feeding $^{15}NH_4Cl$ to both conventional and germ free rats. Conventional rats but not germ free rats incorporated ^{15}N into their body lysine. It was concluded that all of the ^{15}N-lysine was of microbial origin. Metges *et al.* (1999) fed [$^{15}N_2$] urea or $^{15}NH_4Cl$ to human subjects and observed labeling of plasma lysine. Plasma ^{15}N lysine enrichment was greater after intake of $^{15}NH_4Cl$ than [$^{15}N_2$]urea. Hence, humans can absorb microbial synthesized amino acids although the authors were unable to quantitate the net contribution such microbial derived amino acid absorption made to overall amino acid economy (Metges, 2000). Torrallardona *et al.* (2003a) showed that the gastrointestinal microflora in pigs could synthesize a number of essential amino acids using N from fed NH_4Cl, that these amino acids were absorbed from the intestinal tract and that this process made a significant contribution to the amino acid requirements of the host. In pigs, lysine synthesized by the gut microflora is absorbed mainly in the small intestine (Torrallardona *et al.*,

2003b), and as pointed out by these authors there is evidence for a substantial and highly active microbial population in the small intestine of pigs. In summary, there is good evidence in non-ruminants that ammonia released by urea hydrolysis in the gastrointestinal tract can be used by the intestinal microflora to synthesize amino acids, which can be absorbed by the host mainly in the small intestine with some absorption in the large intestine. There is also evidence, at least in pigs, that these microbial synthesized amino acids are quantitatively important to the host. The factors regulating this complex interplay between host and intestinal microflora are unknown, but one potential locus of action for regulating factors could be urea transport proteins. These proteins are an integral component of this N salvage system and are described in the next section.

Urea transport proteins

Urea is a small highly polar molecule that has a low lipid solubility and a low diffusion rate across artificial lipid bilayers. However, urea permeability across cell membranes is actually quite high due to the presence of specific transport proteins. Molecular approaches during the past decade resulted in the cloning of two gene families for facilitated urea transporters, UT-A and UT-B (Sands, 2002). The UT-B gene encodes one isoform, whereas in comparison six UT-A isoforms have been characterized to date (Sands, 2002; Smith *et al.*, 2004). Urea transport proteins have been isolated from a number of tissues as summarized in Table 3. The transporters UT-A1, UT-A2, UT-A3, UT-A4 and the UT-B transporter in the descending vasa recta play a central role in the urinary concentrating mechanism (Sands, 2003; Stewart *et al.*, 2004). The presence of urea transporters in the rumen, small intestine and colon argues that the movement of urea into the gastrointestinal tract is a regulated process. In the case of renal urea transporters, regulating factors have been described (Sands, 1999, 2002 and 2003). For example, a low protein diet increases the abundance of the UT-A1 protein which would result in a decrease in urea excretion (Terris *et al.*, 1998). However there is no information available as to what factors regulate urea transporters in the gastrointestinal tract. Studies performed in ruminants have given conflicting results. In Holstein heifers, an increase in the N content of the diet was associated with an increase in expression of ruminal UT-B, an increase in ruminal ammonia concentration and a decrease in ruminal urease activity (Marini and Van Amburgh, 2003). In lambs, there was no change in UT-B abundance in the rumen, duodenum, ileum or cecum when

Table 3 Urea transport proteins.

Urea transporter	Tissue location	References
UT-A1	Kidney-IMCD*	Sands (2003)
UT-A2	Kidney tDL[†]	Sands (2003)
UT-A2B	Liver Heart	Sands (2003), Klein *et al.* (1999), Duchesne *et al.* (2001)
UT-A3	Kidney-IMCD*	Sands (2003)
UT-A4	Renal medulla[‡] Liver	Sands (2003), Klein *et al.* (1999)
UT-A5	Testis[§]	Sands (2003)
(h)UT-A6	(Human) colon[§]	Smith *et al.* (2004)
UT-A (isoform type?)	(Mouse) colon	Stewart *et al.* (2004)
UT-A (isoform type?)	Brain	Sands (2003)
UT B1/UTB2	Kidney DVR[¶] RBC Brain Multiple tissues including endothelial cells, colon, small intestine, rumen, liver	Sands (2003) Hu *et al.* (2000) Sands (2003), Marini *et al.* (2004)

*Inner medullary collecting duct-kidney.
[†]Thin descending limb-kidney.
[‡]Exact tubular location unknown.
[§]UT-A6 detected in humans only to date; UT-A5 cloned from mouse only.
[¶]Descending vasa recta-kidney.

the N content of the diet was increased (Marini *et al.*, 2004). There was a decrease in ruminal, duodenal, ileal and cecal bacterial urease activity with an increase in the N content of the diet. Clearly more studies need to be done examining the regulation of intestinal urea transporters (abundance and activity).

Protein turnover

Protein turnover involves the balance between protein synthesis and break-down. Since animals cannot store dietary amino acids in excess of the amounts needed for growth and maintenance of protein turnover, then these excess amino acids are preferentially degraded over carbohydrates and lipids. This topic has been reviewed by Waterlow (1999). He puts forward the hypothesis that amino acid supply, in cooperation with insulin, plays a major role in regulating rates of protein synthesis and breakdown in the body.

For the two processes (synthesis and degradation) to be coordinated they must have some factor in common. This common factor is the free amino acid pool, which provides the substrates for synthesis and represents the products of breakdown. Waterlow's model applies to the normal steady state situation (Fig. 4 of Waterlow, 1999). Synthesis is positively correlated with amino acid concentration whereas breakdown is inversely correlated. Balance occurs at the amino acid concentration at which the two curves intersect. One could envisage that each tissue has its own amino acid concentration at which the synthesis and breakdown curves cross and that these tissue differences reflect differences in sensitivity to changes in amino acid concentration or to the action of insulin or other hormones.

In summary, nitrogen balance and protein turnover are interconnected through the free amino acid pool (Fig. 1). Amino acid catabolism liberates ammonia, which is detoxified through the synthesis of urea. Urea production is regulated in the short term by the intrahepatic ammonia concentration and in the long term by changes in the amounts of urea cycle enzymes. A fraction of the urea produced is transferred to the gastrointestinal tract where the urea is hydrolyzed by bacterial urease. A portion of the ammonia so liberated can be used by microbes to synthesize amino acids which can be absorbed by the host. The remainder is absorbed by the host and used either for synthesis of non-essential amino acids or resynthesized into urea.

The movement of urea out of the hepatic parenchymal cell, across the intestinal wall and across the wall of the distal part of the (renal) nephron is facilitated by specific urea transport proteins. The regulation of renal urea transport proteins has been characterized to some extent, but there are very little data with respect to the regulation of hepatic or intestinal urea transporters.

In the normal steady state, the free amino acid pool remains stable because appearances (sum of intake and proteolysis) and disposal (sum of synthesis and catabolism) are equal (Fig. 1). As pointed out by Waterlow (1999) although the achievement and maintenance of N balance is a fact of life, there are many features not understood, principally the control of urea production and excretion to match N intake and the coordination of protein synthesis and breakdown to maintain a relatively constant lean body mass.

Alterations in urea and protein metabolism in chronic renal failure

One of the hallmarks of chronic renal failure is an elevated blood urea concentration. Urea clearance falls as the kidney fails and as a result, blood urea levels will rise unless the individual has significantly reduced N intake. In practice, the blood urea concentration will reflect the balance between the magnitude of the reduction in urea clearance and the level of N intake, assuming the individual is in a steady state and that there is no accompanying protein catabolic process present.

In addition, Walser (1974) has shown that in patients with chronic renal failure, the absolute amount of urea transferred to the intestine is about the same as in normal individuals. However, since blood urea values are higher in patients with renal failure than in normals, the extrarenal (gastrointestinal) clearance of urea is about 3 ℓ/day in patients with chronic renal failure compared to about 18 ℓ/day in normal individuals. This observation is puzzling, as one might have predicted a higher transfer rate of urea across the wall of the intestine in patients with chronic renal failure given their higher blood urea concentrations.

Varcoe *et al.* (1975) studied urea kinetics in a normal group and in patients with chronic renal failure ingesting a diet of either 30 or 70 g protein·d^{-1}. Mean urea production rate was 13.9 mmol/hr in normals and 7.8 mmol/hr and 16.2 mmol/hr in the patients on a 30 g and 70 g protein diet, respectively. Urea hydrolysis in the intestine as a percentage of production was 39% (patients on a 30 g diet), 30% (patients on a 70 g diet) and 17% (normal) group. Mean urea pool size was 932 mmol in the two patient groups combined but only 158 mmol in the control group. Urea half-life was 65 hours in the two patient groups combined but only 8.3 hours in the normal control group. Varcoe *et al.* (1975) also measured the rate of utilization of urea nitrogen (liberated by hydrolysis) for the synthesis of albumin. In the control group about 0.2% of N available from urea degradation was used for albumin synthesis, whereas the value was 1.3% for the two patient groups combined.

The effects of chronic renal failure on the abundance of urea transport proteins have been studied but the results do not give a consistent picture. Rats subjected to a 5/6 nephrectomy displayed a greater than five fold increase in the abundance of hepatic UT-A (Klein *et al.*, 1999). This upregulation of hepatic UT-A was due to the metabolic acidosis present in these uremic animals, since this increase in hepatic UT-A abundance could be prevented by the administration of sodium bicarbonate to these uremic rats (Klein *et al.*, 2002). In addition, normal rats made acidotic by feeding them hydrochloric acid also showed an increase in the abundance of hepatic UT-A (Klein *et al.*, 2002). Hence, acidosis (or perhaps a change in ammonia concentration) rather than uremia *per se* is the signal result-ing in upregulation of UT-A. Upregulation of UT-A in the heart has been demonstrated in rats subjected to a 5/6 nephrectomy, rats made hyper-tensive, and in the myocardium of humans suffering from terminal heart failure secondary to a dilated cardiomyopathy (Duchesne *et al.*, 2001). The authors speculate that the common feature in these three situations is the presence of cardiac hypertrophy. The rats subjected to the 5/6 nephrec-tomy did have evidence of cardiac hypertrophy and hypertension. Urea can be produced as a by-product of ornithine synthesis from arginine via arginase in the first step of the polyamine synthesis pathway. This pathway is present in the heart and polyamine production increases in conditions associated with cardiac hypertrophy. Upregulation of cardiac UT-A may be important for urea exit (from cardiac cells) in conditions where urea pro-duction is increased (Duchesne *et al.*, 2001). Hu *et al.* (2000) reported that rats made uremic by 5/6 nephrectomy displayed a significant decrease in the abundance of renal transporters (UT-A1, UT-A2, UT-B1) and a signifi-cant reduction in the expression of brain UT-B1 mRNA. They were not able to measure UT-B1 protein in the brain due to technical problems. No details are given as to whether the rats were acidotic or hypertensive. Although it is difficult to formulate an internally consistent model based on this data what is clear is that urea movement within the kidney, heart, liver and brain is significantly altered in the setting of chronic renal failure.

A number of studies have been done examining nutritional status and protein turnover in patients with chronic renal failure, but the results are not consistent.

Resting metabolic rate in patients with chronic renal failure has been reported as similar to that of matched controls in several cross-sectional studies (Schneeweiss *et al.*, 1990; Monteon *et al.*, 1986), but decreased com-pared to controls in the cross-sectional study performed by O'Sullivan *et al.* (2002). Panesar and Agarwal (2003) using regression analysis described a fall in resting metabolic rate with a decrease in GFR, whereas Kuhlmann *et al.* (2001) in a longitudinal study reported an increase in resting metabolic rate with declining GFR.

In a large cross-sectional study (Kopple *et al.*, 2000), baseline measure-ments in patients with chronic renal failure showed that dietary protein and energy intake and serum and anthropometric measures of protein-energy nutritional status progressively declined as the GFR decreased. Protein-energy malnutrition is a common occurrence in patients with chronic renal failure before the start of dialysis (Lindholm *et al.*, 2002). One factor thought to be important in the genesis of protein-energy malnutrition is chronic acidosis (Lindholm *et al.*, 2002; O'Sullivan *et al.*, 2002). O'Sullivan *et al.* (2002) observed that patients with only moderate chronic renal failure had significantly reduced lean body mass and bone mineral content compared to matched controls. The patients with chronic renal failure studied by O'Sullivan *et al.* (2002) were more acidotic than the control group.

Whole body protein turnover has also been studied in patients with chronic renal failure. Protein turnover is determined from the kinetics of infused carbon labeled leucine. Total leucine turnover or flux $(Q) = S+O = B + I$ where S is the rate of incorporation of leucine into body proteins, O is the rate of leucine oxidation, B is the rate of leucine appearance from protein degradation and I is the dietary intake of leucine. When subjects are in the postabsorptive (fasting state), leucine intake is zero and total leucine flux (Q) becomes equal to B. The assumption is made that leucine kinetics are representative of amino acid kinetics in general.

Biolo *et al.* (1998) measured whole body leucine kinetics in patients with chronic renal failure after an overnight fast. Most of these patients were treated with calcium carbonate so that blood pH and serum calcium levels were close to normal values. They found a positive linear relationship between the values of plasma creatinine concentration and the rate of whole body protein turnover. This correlation suggests that worsening renal function is associated with accelerated rates of whole body protein turnover and is consistent with the previously noted observation of Kuhlmann *et al.*, (2001) that resting metabolic rate increases with declining GFR. Since these studies were done in fasting patients, total leucine flux is equal to the rate of protein breakdown. These authors did not report rates of protein synthesis or rates of amino acid oxidation for their patients.

As discussed by Biolo (2001) the genesis of abnormalities of protein metabolism in chronic renal failure is multifactorial. Some conditions such as a low protein-energy intake tend to decrease protein turnover whereas metabolic acidosis, and secondary hyperparathyroidism tend to increase protein turnover. However, in the aforementioned study of Biolo *et al.* (1998) there was no correlation between whole body protein turnover and blood pH, bicarbonate or calcium concentrations.

Castellino *et al.* (1992) reported that patients with chronic renal failure compared to a control group had reduced whole body protein turnover, a reduced rate of protein synthesis, and a reduced rate of amino acid oxidation. Studies were done after an overnight fast. In addition, patients also displayed a reduced rate of protein synthesis with an amino acid infusion compared to the control group. Mean arterial pH and mean arterial bicarbonate concentration were normal in the patients with chronic renal failure. Adey *et al.* (2000) on the other hand found whole body protein turnover, the rate of protein synthesis and amino acid oxidation to be normal in patients with chronic renal failure compared to a control group. Studies were done after an overnight fast. The patients did display, however, a decreased rate of synthesis of mixed muscle protein, myosin heavy chain and muscle mitochondrial protein compared to controls. In addition, muscle mitochondrial enzyme activity was also significantly lower in patients with chronic renal failure than in the control group. These authors found, based on regression analysis, that the rate of synthesis of myosin heavy chain, mixed muscle protein and muscle mitochondrial enzymes were positively correlated with GFR. This correlation suggests that as GFR decreases the rates of synthesis for muscle proteins and muscle mitochondrial enzymes also decreases. Mean serum CO_2 (a measure of acid-base balance) in the patients with chronic renal failure was normal.

Goodship *et al.* (1990) measured the effects of a varied protein intake as well as fasting and feeding on protein metabolism in patients with chronic renal failure as well as a matched control group. Whole body protein turnover as well as protein synthetic and degradation rates were the same in the chronic renal failure and control groups.

In all of these studies, except for the one by Goodship *et al.* (1990) acid-base status was noted to be normal. There is no comment regarding the acid-base status of the patients in the paper by Goodship *et al.* In two of the studies (Biolo *et al.*, 1998; Castellino *et al.*, 1992), patients were noted to be taking medication to prevent a metabolic acidosis. Thus the changes in whole body protein turnover and in the rates of protein synthesis described in these studies are not due to the presence of a metabolic acidosis but are the result of some other feature of the uremic milieu.

The protein kinetic data are consistent with the following model. Early in the course of renal failure, whole body protein turnover is either normal or decreased but the rate of protein synthesis is reduced (Adey *et al.*, 2000; Castellino *et al.*, 1992). As kidney function progressively declines, whole body protein turnover increases (Biolo *et al.*, 1998) while the rate of protein synthesis decreases (Adey *et al.*, 2000). Hence, the increasing whole body protein turnover with advancing renal failure represents a catabolic state characterized by accelerated protein breakdown. This model is consistent with Kuhlmann *et al.'s* finding that resting metabolic rate increased with a reduction in GFR. As noted in the "Metabolic Rate and Renal Function" section, in the normal mammal whole body protein turnover accounts for about 20%–30% of resting metabolic rate. Guarnieri *et al.* (2003) have recently reviewed the topic of malnutrition and protein metabolism in patients with chronic renal failure. They propose that "uncomplicated" uremic patients have decreased whole body protein turnover as well as decreased rates of both protein synthesis and degradation. Patients with more advanced uremia display increased whole body protein turnover and increased protein degradation because of the development of a metabolic acidosis or a superimposed complication such as an acute or chronic infection. However, as already noted, the data of Biolo *et al.* (1998) showing a correlation between whole body protein turnover and serum creatinine was derived from patients with controlled acid-base status. In addition, understanding the abnormalities of protein metabolism in the setting of chronic renal failure is made more difficult by the many unanswered questions in the normal human (Waterlow, 1999; Liu and Barrett, 2002). For example, one major question concerns the nature of the coupling between the processes of protein synthesis and protein degradation.

Additional Aspects of Renal Failure

The major manifestations of the uremic state arise from the complex interplay between decreased renal excretory function and abnormalities in N and protein metabolism. The study results of Nelson *et al.* (1985) briefly described in the "Introduction" section support this contention. These authors successfully managed 15 patients with severe renal failure for a number of years primarily with a low protein (0.38 g/kg body weight) but calorically adequate diet. The only medications used, when indicated, were calcium supplements, phosphate binders, uric acid lowering agents, and antihypertensive drugs. Body composition using dilution of the isotope D_2O in total body water was performed in five patients before and while they were on the diet. Lean body mass was 51.8 kg at the start and 50.6 kg after ten months on the diet. Mean serum albumin concentration increased in these patients. The maintenance of lean body mass and the increase in serum albumin concentration indicates that the low protein diet in these patients reversed the catabolic state observed in advanced renal failure.

In fact the study of Goodship *et al.* (1990) previously discussed supports this conclusion. In their study when patients with chronic renal failure were switched from a high protein (1.0 g/kg/day) to a low protein (0.6 g/kg/day) diet the rate of amino acid oxidation was reduced. A low protein diet can improve many of the manifestations of uremia (Aparicio *et al.*, 2001) and substantially delay the date of initiation of renal replacement therapy (Nelson *et al.*, 1985; Aparicio *et al.*, 2001).

However, there are other aspects of chronic renal failure that need to be considered. A chronic metabolic acidosis occurs as renal failure progresses. The primary factor responsible for the genesis of the acidosis is a decrease in renal ammonia production as nephron mass is reduced (Krapf *et al.*, 2000). Chronic metabolic acidosis leads to a significant increase in protein breakdown, protein synthesis and the rate of amino acid oxidation consistent with a predominantly catabolic state (Reaich *et al.*, 1992). Correction of the metabolic acidosis with sodium bicarbonate reverses these abnormalities (Reaich *et al.*, 1993). There are a number of other studies demonstrating that metabolic acidosis stimulates the catabolism of essential amino acids in muscle and activates the ubiquitin-proteasome proteolytic system (Bailey and Mitch, 2004). Metabolic acidosis has other multiple effects on the action of various hormones and on bone turnover (Bailey and Mitch, 2004). Patients with chronic renal failure also suffer from skeletal abnormalities collectively known as renal osteodystrophy. One component of this collection of abnormalities is due to secondary hyperparathyroidism. This topic is discussed in Chapter 4. Cardiovascular disease, particularly ischemic heart disease and left ventricular hypertrophy has a much higher incidence in patients with chronic renal failure (McMahon and Parfrey, 2004). The pathogenesis of cardiovascular disease in patients with chronic renal failure involves both so called traditional risk factors and uremia-related risk factors (McMahon and Parfrey, 2004). The latter include: platelet dysfunction, hyperhomocysteinemia, abnormal divalent ion metabolism. The cardiovascular complications of renal failure will not be discussed any further but the topic of atherosclerotic vascular disease is reviewed in Chapter 3.

Natural Animal Model

The American black bear (*Ursus americanus*) survives winter for up to five months at near-normal body temperature without eating, drinking, urinating or defecating (Singer, 2002; Nelson *et al.*, 1973). In essence, the bear's physiological situation can be characterized as a combination of renal failure and fasting. When the bear enters its den and goes into winter sleep, its body temperature decreases by only a few degrees from about 37°C–39°C to 34° (Nelson *et al.*, 1973; Brown *et al.*, 1971). Brown *et al.* (1971) measured rectal temperatures in six captive black bears in both the active and dormant state. In the active state mean rectal temperature was 36.9°C and in the dormant state 34.8°C. During the dormant period, resting metabolic rate decreases by about 50% (Nelson, 1989; Hellgren, 1998) and GFR decreases by about 60% (Brown *et al.*, 1971). The reduced GFR during the dormant state is the result of hemodynamic alterations since effective renal plasma flow (based upon clearance of p-aminohippuric acid) and cardiac output are decreased by 57% and 38%, respectively during dormancy. However, the bear is not a deep hibernator and during winter sleep the bear can arouse itself quickly into a mobile reactive state aware of its surroundings (Nelson *et al.*, 1973). Nelson *et al.* (1984) have established that the bear begins preparing for winter denning in late summer and early fall while food is available. Biochemical markers of the state of hibernation may be present weeks before the bear actually dens. In fact, Nelson *et al.* (1975) were not able to induce the metabolic adaptations characteristic of the dormant state, when bears were deprived of food and water in the summer but housed in a darkened and cold environment.

Even though the bear does not eat, drink, urinate or defecate during dormancy and even though urea production continues albeit at a reduced rate, blood urea concentration actually falls during this dormant period (Barboza *et al.*, 1997; Wright *et al.*, 1999; Nelson *et al.*, 1975).

The concentration of uric acid, another nitrogenous end product, does not change during the denning period compared to the active state (Wright *et al.*, 1999).

Barboza *et al.* (1997) studied whole body urea cycling and protein turnover during autumn hyperphagia and winter dormancy in captive bears. Mean plasma urea concentration fell from 54.2 mg/dℓ in the active state to 13.5 mg/dℓ during mid-dormancy. This decline in plasma urea concentration was not accompanied by a statistically significant change in urea pool size. However, turnover time for the urea pool was much longer in the winter than in the autumn. In the autumn average urea N production was 27.5 mmol.kg$^{-0.75}$.d^{-1}, and in the winter 4.7 mmol.kg$^{-0.75}$.d^{-1}; a reduction of 83%. Most of the urea N produced in the autumn was excreted with only about 7.8 mmol.kg$^{-0.75}$.d^{-1} recycled to the intestine. Only 1.1% of the recycled urea N was reutilized for amino acid synthesis. In contrast, almost

100% of urea N produced during dormancy was recycled to the intestine and essentially all of the recycled urea N was used for amino acid synthesis. Turnover time for urea was 5.1 hours in the active state and 43.0 hours during denning. Calculated urea half-life is about 3.5 hours in the active state and about 30 hours in the dormant state. Nelson *et al.* (1975) measured urea half-life in two bears and reported values of nine and 12 hours prior to denning and 25 hours and 90 hours during dormancy. In these two bears blood urea concentration decreased slightly as did urea pool size. Protein turnover measurements conducted by Barboza *et al.*, (1997) were consistent with urea kinetics. In the autumn, protein turnover was 21.5 $g.kg^{-0.75}.d^{-1}$ with a rate of synthesis of 9.3 $g.kg^{-0.75}.d^{-1}$ and a rate of amino acid oxidation of 12.2 $g.kg^{-0.75}.d^{-1}$. During the dormant period, protein turnover was reduced to 15.2 $g.kg^{-0.75}.d^{-1}$ and the rate of synthesis increased to 14.4 $g.kg^{-0.75}.d^{-1}$; not statistically different from the autumn rate. However, the rate of amino acid oxidation was significantly reduced to 0.8 $g.kg^{-0.75}.d^{-1}$; a reduction of 93%. The reduction in amino acid oxidation was mirrored by reductions in the circulating levels of aminotransferases. During autumn hyperphagia body mass increased with most of the increase due to accumulation of fat. During dormancy most of the mass lost was again fat since about 92% of energy expended in the dormant state was derived from fat oxidation. Lean body mass is well preserved during dormancy (Barboza *et al.*, 1997; Nelson *et al.*, 1975; Lundberg *et al.*, 1976).

In non-lactating females, plasma ammonia levels were lower during denning (116 μM) compared to the active state (182 μM) (Wright *et al.*, 1999). This difference is probably due to the significant suppression of amino acid oxidation during dormancy. As a marker of renal failure, plasma creatinine levels increase by two to three fold between active and dormant states (Barboza *et al.*, 1997; Wright *et al.*, 1999). Since GFR is about 40% of that of the active state, urea and urine are excreted into the urinary bladder; about 4 g of urea and 100 ml of urine each 24 hours (Nelson *et al.*, 1973). However, the urinary bladder of the bear is capable of reabsorbing this amount of urea and water each day (Nelson *et al.*, 1975). Hence while dormant, the bear maintains its pre-denning rate of protein synthesis, while reducing amino acid oxidation by over 90%. The small amount of urea produced is completely degraded in the intestine and essentially all of the urea N is reutilized for amino acid synthesis. The urinary bladder plays a central role in maintaining the state of winter sleep by absorbing water and urea at a rate equal to their entry into the urinary bladder. However, water and urea can also be absorbed by the canine urinary bladder especially under conditions of low urine flow (Levinsky and Berliner, 1959). Although this picture in the bear is consistent with the data discussed, Wolfe *et al.* (1982) have presented experimental evidence that the recycling of urea N into amino acids in the bear may not occur exclusively via intestinal urea catabolism and that other (unknown) pathways may exist. How do these adaptations in urea metabolism and protein turnover during the bear's denning period compare to adaptations observed in other mammals under comparable circumstances.

Owen *et al.* (1969) studied obese subjects who fasted for five weeks. Daily intake during starvation consisted of a multivitamin capsule, 17 meq of NaCl, 1500 ml of water and intermittently 13 meq of KCl. Liver and kidney metabolism were carefully studied during this period. Although urea kinetics were not measured, urea production and hydrolysis can be estimated by the collected data. After five to six weeks of starvation, mean hepatic blood flow was $1.08 \, \ell/min$ and mean splanchnic arterial-venous difference for alpha amino nitrogen was 0.14 mmoles/ℓ. It is assumed that this difference represents amino acids catabolized in the liver and that all of the alpha amino nitrogen is incorporated into urea. Using these assumptions, $3.05 \, g$ urea nitrogen (218 mmoles urea nitrogen) would be produced per day. The average weight of these subjects at the end of the fast was $115 \, kg$. Using the allometric equation for urea production (Table 1), an individual of this weight with a "normal" diet would have a predicted urea production rate of $19.1 \, g$ urea N/day. Hence these individuals suppressed urea production by about 84%. Mean urea excretion at the end of the five to six weeks was $1.55 \, g$ urea nitrogen/day. Hence urea kinetic parameters (in g urea N/day) in this group of subjects would be as follows: urea production $= 3.05$, excretion $= 1.55$ (51% of production), and intestinal hydrolysis $= 1.5$ (49% of production). Hence these subjects were capable of suppressing urea production (and presumably amino acid oxidation) as effectively as the bear, but were much less efficient in recycling urea nitrogen than the bear. However Holstein heifers on a low N intake (as described in the "Nitrogen and Urea Metabolism" section) did recycle 83% of produced urea N into the intestine. Motil *et al.* (1981) examined whole body leucine metabolism in young men consuming a dietary protein intake $(g.kg^{-1}.d^{-1})$ of either 1.5, 0.6 or 0.1. In the fed state, when subjects were switched from the 1.5 to the 0.1 protein diet, the absolute rates of whole body leucine flux (protein turnover), leucine incorporation into protein (protein synthesis) and leucine oxidation fell by 54%, 43% and 75%, respectively.

Hence other mammals, when switched to a zero or low nitrogen intake can suppress urea production and amino acid oxidation almost as well as the bear. However other mammals cannot recycle urea nitrogen to the intestine to the same extent as the dormant bear nor can other mammals match the dormant bear's ability to utilize all of the recycled urea nitrogen for protein synthesis. Other mammals when subjected to a restricted protein intake have a decreased rate of protein synthesis. As reviewed by Waterlow (1984), total fasting in humans produces a decrease in whole body protein synthesis of about 15% to 27%. However these fasts were only of one to two weeks in duration. In rats fasted for 72 hours, muscle protein synthesis falls by about 50% compared to a control fed group (Li *et al.*, 1979). Fasting in humans also reduces the rate of protein degradation. Young *et al.* (1973) studied muscle protein catabolism in three obese subjects fasted for 20 days. The rate of catabolism progressively fell and by day 20 of the fast was 34% less than on day 3. As pointed out by Owen *et al.* (1969) and Cahill (1970), a reduction in protein catabolism during prolonged starvation is mandatory for survival in humans. Without such an adaptation, total body protein would be rapidly depleted (Cahill, 1970). In contrast, the hibernating bear, even though it has no intake, behaves quite differently from the fasting human. During dormancy the (fasting) bear's rate of protein synthesis is equivalent to or greater than that of the active (feeding) bear, and the rate of protein degradation increases in the dormant (fasting) bear (Lundberg *et al.*, 1976; Barboza *et al.*, 1997).

Other studies in the bear support and amplify these differences between the bear and other mammals. Lundberg *et al.* (1976) examined protein metabolism in bears by measuring the fate of labeled albumin and leucine following intravenous injections. The albumin used in these studies was derived from the bears used in these experiments. The albumin was non-antigenic in the experimental animals. These investigators measured the rate of labeled albumin disappearance from the plasma and the rate and extent of labeled leucine incorporation into plasma proteins in bears in the active state and in these same bears in the dormant state. The rate of albumin turnover was about three fold higher when the bear was dormant as compared to the active state. Since serum albumin concentration was constant between the two states, the increased turnover reflected an equal increase in both albumin synthesis and degradation. Following injection of labeled leucine, this amino acid appeared in plasma proteins in increased quantities during dormancy compared to the active state. The investigators proposed that the increased uptake of leucine during dormancy had to be due to an increased rate of protein synthesis because plasma leucine and protein concentrations were similar in the active and dormant states.

As pointed out by Wright *et al.* (1999) a decrease in essential, relative to non-essential, plasma amino acids is seen in prolonged fasting in humans. Yet the dormant bear is able to maintain the plasma concentration of essential amino acids despite many months of fasting in the den (Wright *et al.*, 1999; Nelson *et al.*, 1973). The number of enzymes required by prokaryotic cells to synthesize the nutritionally essential amino acids is large relative to the number of enzymes required to synthesize the nutritionally non-essential amino acids (Rodwell, 2000a). This suggests there is a survival advantage in retaining the ability to manufacture "easy" amino acids while losing the ability to make "difficult" amino acids. The observation that plasma levels of essential amino acids are maintained during dormancy raises the possibility that the bear has the capability of synthesizing essential amino acids. Nelson (1989) has discussed the evidence for this possibility. When labeled leucine is injected during the dormant period, most is quickly cleared from the plasma reappearing in plasma proteins. The data showed continual turnover of leucine in plasma proteins with in addition slow oxidation of leucine. Although plasma leucine is oxidized in winter when the bear is not eating its constant level in blood plasma (Nelson *et al.*, 1973) suggests that the bear may synthesize this essential amino acid *de novo* during dormancy. A similar picture was found for the essential amino acid threonine (Nelson, 1989; Meredith *et al.*, 1988). Labeled threonine was incorporated into plasma proteins and also slowly oxidized. However, its plasma concentration did not change (Nelson *et al.*, 1973; Wright *et al.*, 1999). The data suggest that threonine was being synthesized by the bear. However there are other possible explanations for this data, which do not invoke synthesis of essential amino acids by the bear. Tinker *et al.* (1998) have measured a significant decrease in skeletal muscle protein content during hibernation in both the gastrocnemius and biceps femoris muscle. Tinker *et al.* (1998) estimated that the losses in the gastrocnemius and biceps femoris muscles would represent a whole-body seasonal loss of about 2.2 and 5.2 kg of protein, respectively. Could this degradation of a small but significant amount of skeletal muscle protein be a source of essential amino acids for the bear? These observations of Tinker *et al.* (1998) however, are at odds with the data of Koebel *et al.* (1991). These investigators reported only minimal muscle loss which they estimated to be about 10%–20% over the average six-month denning period. Another possibility is that essential amino acids are synthesized by the intestinal microflora of the bear and then subsequently absorbed by the host. The evidence for this sequence of steps in non-ruminant mammals was reviewed in the "Nitrogen and Urea Metabolism" section.

Table 4 summarizes the main features of urea and protein metabolism in dormant bears, other mammals either fasting or on a low nitrogen intake and humans with chronic renal failure.

Table 4 Urea and protein metabolism in the dormant bear and other mammals.

Parameter	Bear dormant	Other mammals fasting or low N intake	CRF minimal to modest protein restriction
Urea production	D (Barboza *et al.*, 1997)	D (Owen *et al.*, 1969; Marini and Van Amburgh, 2003)	U or D (Varcoe *et al.*, 1975)
Urea hydrolysis as a percent of production	100% (Barboza *et al.*, 1997)	49–92% (Owen *et al.*, 1969; Singer, 2003b; Marini and Van Amburgh, 2003)	30–39% (Varcoe, *et al.*, 1975)
Urea blood concentration	D (Barboza *et al.*, 1997; Wright *et al.*, 1999; Nelson *et al.*, 1975)	D (Marini and Van Amburgh, 2003)	I (Varcoe *et al.*, 1975)
Urea pool size	U or slight D (Barboza *et al.*, 1997; Nelson *et al.*, 1975)	D (Marini and Van Amburgh, 2003)	I (Varcoe *et al.*, 1975)
Urea half life	I (Barboza *et al.*, 1997; Nelson *et al.*, 1975)	D (Marini and Van Amburgh, 2003)	I (Varcoe *et al.*, 1975)
Percent of urea N (hydrolysis) used for amino acid synthesis	100% (Barboza *et al.*, 1997)	0.2–60% (Singer, 2003b; Marini and Van Amburgh, 2003; Varcoe *et al.*, 1975)	1.3% (Varcoe *et al.*, 1975)
Urea reabsorption from urinary bladder	Yes (Nelson *et al.*, 1975)	Yes (Levinsky and Berliner, 1959)	?

Table 4 (*Continued*).

Parameter	Bear dormant	Other mammals fasting or low N intake	CRF minimal to modest protein restriction
Protein turnover	Slight D (Barboza *et al.*, 1997)	D (Motil *et al.*, 1981)	U or D (early) (Castellino *et al.*, 1992; Adey *et al.*, 2000; Goodship *et al.*, 1990) I (late) (Biolo *et al.*, 1998)
Rate of protein synthesis	U or I (Barboza *et al.*, 1997; Lundberg *et al.*, 1976)	D (Motil *et al.*, 1981)	U or D (Castellino *et al.*, 1992; Adey *et al.*, 2000; Goodship *et al.*, 1990)
Rate of amino acid oxidation	D (Barboza *et al.*, 1997)	D (Motil *et al.*, 1981)	U or D (Castellino *et al.*, 1992; Adey *et al.*, 2000; Goodship *et al.*, 1990)

D = decreased, I = increased, U = unchanged.
CRF: Chronic renal failure.

Bear and Small Mammal Hibernators

There are significant differences between the characteristics of hibernation in the bear and in small mammals which comprise the majority of mammalian hibernators (Heldmaier *et al.*, 2004). During hibernation the bear does not exhibit any spontaneous arousals (Harlow *et al.*, 2004), reduces its core temperature by only a few degrees (Hellgren, 1998), is able to maintain a metabolic rate about 50% of resting metabolic rate (Hellgren, 1998) and an overall rate of protein synthesis equal to or greater than that of the active state (Barboza *et al.*, 1997). By contrast, small mammal hibernators display periodic arousals during the hibernation season, generally reduce their core temperature and metabolic rate substantially and demonstrate a marked suppression of overall protein synthesis. For example, marmots have a hibernation period beginning in late September and lasting through the end of March or early April (Heldmaier *et al.*, 2004). However this period actually consists of a series of hibernation bouts. Each bout is characterized by four different states: entrance into hibernation, maintenance of deep hibernation (torpor) for a number of days, which is then terminated by an arousal stage, and followed by a euthermic period of several days. Thereafter the cycle is repeated. In the case of the marmot, hibernation bouts last between 12 and 20 days (Heldmaier *et al.*, 2004). In the big brown bat (Yacoe, 1983), the mean duration of the hibernation (torpid) state was 155 hours with a range of 17 to 471 hours. Mean duration of spontaneous arousals was 3.2 hours with a range of 0.5 to 11.5 hours. During the hibernation season, the big brown bat spends 98% of overall time in torpor and 2% in arousals. During the phase of deep hibernation, there is a marked depression of metabolic rate as well as a large reduction in the overall rate of protein synthesis, while at the same time allowing synthesis of selected key proteins (Storey, 2003). The marmot reduces its core temperature from 36°C to 13°C and its metabolic rate from 0.34 ml $O_2/g/hr$ to 0.014 ml $O_2/g/hr$ during a hibernation bout (Heldmaier *et al.* 2004). The major metabolic cost during the hibernation season occurs during arousals and subsequent euthermic periods. In the marmot (Heldmaier *et al.*, 2004), 74% of all energy reserves required for the entire hibernation season are consumed during arousals (17%) and euthermic periods (57%). In the big brown bat (Yacoe, 1983), arousals account for only 2% of overall time in the hibernation season, but over 80% of the metabolic expenditure.

Jaroslow *et al.* (1968) measured the *in vivo* incorporation of leucine into serum proteins in hibernating (torpid) and active ground squirrels. The squirrels were given an intraperitoneal injection of radioactively labeled leucine and 17 hours later, the radioactivity of serum proteins (albumin and gamma globulin) was measured. Synthesis of serum albumin and gamma globulin was reduced by 75%–94% in hibernating compared to active squirrels. Frerichs *et al.* (1998) measured *in vivo* rates of leucine incorporation into protein in active and torpid 13-lined ground squirrels. In brain, liver and heart, rates of protein synthesis in hibernating squirrels were between 0.03% and 0.10% of the mean rates measured in active squirrels. This suppression of protein synthesis during torpor was not entirely a passive thermodynamic consequence of the lowered core temperature since brain extracts from hibernating squirrels studied at 37°C *in vitro* incorporated leucine into protein at a rate only 30% of that of brain extracts from active squirrels. In the big brown bat, the mean fractional rates of protein synthesis measured *in vivo* as incorporation of radioactively labeled phenylalanine into protein in the pectoralis muscle and liver were extremely low (and in fact below the limit of detection of the technique used) during hibernation (torpid) states but significantly higher during arousals (Yacoe, 1983). In fact, the mean fractional rates of hepatic protein synthesis during arousals and in summer active bats did not differ. However, for the pectoralis muscle, the mean fractional rate of protein synthesis during arousals was about 25% of the rate in summer active bats. The suppression of protein synthesis during torpor appears to involve regulation of mRNA translation by the ribosomal machinery (Frerichs et al., 1998). The key to translational suppression is reversible phosphorylation control over the activities of ribosomal initiation and elongation factors (Storey, 2003). For example, phosphorylation of the alpha subunit of the eukaryotic initiation factor 2 inhibits the function of this factor. The content of the phosphorylated initiation factor is higher in the kidney and brain of torpid versus euthermic 13-lined ground squirrels (Storey, 2003). A similar finding has been measured in multiple tissues of the torpid bat Myotis lucifugus compared to the active euthermic bat (Storey, 2003).

Hibernation is most frequently observed in mammalian orders with species of small size and is largely lacking in orders with large species. As already noted, for these small mammals hibernation is not continuous throughout the season but consists of multiple bouts of deep torpor lasting one to three weeks interspersed with brief arousals of up to a day. Protein synthesis is one of the major energy expenditures of cells, requiring about five ATP equivalents per peptide bond formed and consuming as much as 40% of the total ATP turnover in selected organs such as liver (Storey, 2003). Hence it is not surprising that substantial suppression of the overall rate of protein synthesis is an integral component of the metabolic rate depression during hibernation in these small mammals. However, as measured in the big brown bat, significant rates of protein synthesis do occur during periods of arousal. The bear is probably the largest mammal to hibernate (Heldmaier et al., 2004) and in this mammal which reduces metabolic rate by about 50% during hibernation, the lack of spontaneous arousals and the maintenance of a rate of protein synthesis at least equal to that of the active state are key features underlying its success as a hibernator.

Concluding Remarks and Future Directions

When the bear enters its dormant period a number of metabolic/physiological adaptations occur. The rate of amino acid oxidation is profoundly reduced whereas the rate of protein turnover remains about the same as that prior to dormancy. The rate of protein synthesis is either similar to or higher than the predormancy rate depending upon which data are used. This combination of a very low rate of amino acid oxidation with a maintained or increased rate of protein synthesis separates the dormant bear from other mammals ingesting a low nitrogen diet. Motil *et al.* (1981) clearly demonstrated that in humans, lowering the daily protein intake from about 100 g to 7 g resulted in a decrease in the rate of amino acid oxidation (like the bear) but also a significant decrease in the rate of protein synthesis (unlike the bear). In the case of patients with renal failure, the data generally show that the rate of protein synthesis is decreased, as is the rate of amino acid oxidation compared to controls. Quantitatively the bear is also different. Dormant bears have a rate of amino acid oxidation only 7% of the rate observed in the active state, whereas the subjects of Motil *et al.* (1981) on a protein diet of 7 g per day had a rate of amino acid oxidation 25% of the rate in these same subjects when on a 100 g daily protein intake.

During dormancy, amino acid oxidation, although at a low rate, results in the continuous production of urea. Although the bear does not urinate or defecate, blood urea concentration actually decreases despite this continuous urea synthesis. The dormant bear maintains a GFR about 30%–40% of normal so that a fraction of the produced urea is excreted in the urine. However this urea (and urine) is completely reabsorbed by the urinary bladder. Essentially all of the produced urea is disposed through its transfer to the intestinal tract with subsequent hydrolysis by microbial urease. However, Wolfe *et al.* (1982) has presented evidence that the recycling of urea nitrogen into amino acids may not occur exclusively via intestinal urea catabolism and that other (unknown) pathways may exist. Current data indicate that the dormant bear utilizes essentially all of the recycled urea nitrogen for protein synthesis — a feat not matched by other mammals as summarized in Table 4. While dormant, the bear does not eat or drink, continues to oxidize amino acids albeit at a low rate and yet plasma concentrations of essential amino acids are well maintained (Wright *et al.*, 1999; Nelson *et al.*, 1973). This observation raises the intriguing possibility that the bear, unlike other mammals, synthesizes essential amino acids. This possibility has already been discussed in the "Natural Animal Model" section. Nelson (1989) has proposed that the primary development of formation of essential amino acids from urea nitrogen may well represent an evolutionary event necessary for the bear to survive in winter. However the microflora of the mammalian intestinal tract (small and large intestine) is capable of synthesizing all of the essential amino acids (Morrison and Mackie, 1997; Torrallardona *et al.*, 2003a). In addition, as discussed in the "Nitrogen and Urea Metabolism" section, ruminants and non-ruminants can absorb microbial synthesized essential amino acids and that these microbial derived essential amino acids make important contributions to the amino acid requirements of the host mammal. If indeed the bear does synthesize essential amino acids, this would be a unique adaptation as suggested by Nelson (1989). However, based on current data a more reasonable approximation is that the dormant bear relies on its intestinal microflora as the source of essential amino acids. Synthesis of essential amino acids by gut microbes is a mechanism already in place in mammals. Furthermore, this approximation fits in with other aspects of urea and protein metabolism in the bear. Almost all mammals recycle urea nitrogen to the intestine and reutilize a variable fraction of this urea nitrogen for protein synthesis (Singer, 2003b). The dormant bear has quantitatively extended these adaptations; recycling essentially all of the urea nitrogen to the intestine and reutilizing all of this urea nitrogen for protein synthesis. However, there remain many unanswered questions concerning urea and protein metabolism in dormant bears. However, before discussing this topic, what the dormant bear's known adaptations mean with respect to the management of chronic renal failure in humans is reviewed.

In the "Natural Animal Model" section, the physiological situation of the dormant bear can be characterized as a combination of renal failure and fasting. Yet despite this physiological combination, the bear remains well during the dormant period. As discussed in the "Alterations in Urea and Protein Metabolism in Chronic Renal Failure" section, abnormal urea and protein metabolism plays a central role in the genesis of the uremic syndrome. The observation that the bear's major adaptations in the dormant state involve changes in urea and protein metabolism affirms this central role. Furthermore, the adaptations of the bear to dormancy appear to be essentially extensions of adaptations that most mammals make to a restricted nitrogen intake. An obvious exception to this concept is the possibility that the dormant bear synthesizes essential amino acids. If the bear's adaptations could be induced in a human with chronic renal failure, then the point of time at which dialysis had to be initiated could be delayed significantly. Nelson *et al.* (1985) (see "Introduction" section) demonstrated that patients with severe renal failure ingesting a low amount (0.38 g/kg/day) of high biological value protein could be maintained in a well state for a prolonged time period. Induction of the bear's adaptations in patients already receiving dialysis would allow these patients to remain well with less frequent dialysis treatments. The only indication for dialysis in these patients would be for maintenance of volume and electrolyte balance. Patients with end-stage kidney disease who continued to have a good urine output would not require dialysis at all since the metabolic derangements of the uremic state would be corrected by the induced bear adaptations. However, using the bear's adaptations for therapeutic interventions would require much more information about these adaptations.

The number of unanswered questions, with respect to urea and protein metabolism in the bear and with respect to urea and protein metabolism in general, are many. Some of these research questions are as follows:

(1) What events occur during the predormant state that allow the bear to make the appropriate adaptations during dormancy? Are there specific substance(s) that trigger or induce the dormant state?

(2) The bear most certainly has urea transport proteins as present in other mammals, although the actual presence of these urea transporters has not been documented in the bear. This family of urea transporters clearly forms a coordinated system for regulating urea metabolism. However, the factors regulating urea transporters (except for some data with respect to renal urea transporters) are essentially unknown. In the bear, urea transporters in the liver, intestine, kidney and urinary bladder would allow for coordination of urea movement out of the liver and across the walls of the intestine, renal tubules, and urinary bladder. The bear could regulate the extent of urea recycling to the intestine by changing the function of these transporters and hence allow for essentially complete recycling in the dormant state. Currently there are no data to support this concept.

(3) Does the gut microflora of the bear differ from that of other mammals? How is the gut microflora nutritionally supported when the bear is dormant and not eating? Is the dormant bear more efficient in reabsorbing microbial synthesized essential amino acids than other mammals or than the bear when in the active state?

(4) Does the bear actually synthesize essential amino acids while dormant?

These are but a few of the questions that would need to be answered before one could consider trying to "transfer" the dormant bear's adaptations to humans with chronic renal failure. However, the bear teaches us that a biological solution to the problem of chronic renal failure exists although nature is very far ahead of us in designing these solutions.

References

Adey, D., Kumar, R., McCarthy, J.T., Nair, K.S., 2000. Reduced synthesis of muscle proteins in chronic renal failure. *Am. J. Physiol. Endocrinol. Metab.* 278, E219–E225.

Aparicio, M., Chauveau, P., Combe, C., 2001. Are supplemented low-protein diets nutritionally safe? *Am. J. Kidney Dis.* 37, Suppl. 2 (January), S71–S76.

Bailey, J.L., Mitch, W.E., 2004. Pathophysiology of uremia. In: Brenner, B.M. (Ed.), *Brenner and Rector's the Kidney*, 7th Edn. Saunders, Philadelphia, PA, pp. 2139–2164.

Barboza, P.S., Farley, S.D., Robbins, C.T., 1997. Whole-body urea cycling and protein turnover during hyperphagia and dormancy in growing bears (*Ursus americanus* and *U. arctos*). *Can. J. Zool.* 75, 2129–2136.

Biolo, G., Toigo, G., Ciocchi, B., Morena, G., Situlin, R., Vasile, A., Carraro, M., Faccini, L., Guarnieri, G., 1998. Relationship between whole-body protein turnover and serum creatinine in chronically uremic patients. *Miner. Electrolyte Metab.* 24, 267–272.

Biolo, G., 2001. Can we increase protein synthesis by anabolic factors? *Am. J. Kidney Dis.* 37, Suppl. 2 (January), S115–S118.

Bricker, N.S., 1969. On the meaning of the intact nephron hypothesis. *Am. J. Med.* 46, 1–11.

Brown, D.C., Mulhausen, R.O., Andrew, D.J., Seal, U.S., 1971. Renal function in anesthetized dormant and active bears. *Am. J. Physiol.* 220, 293–298.

Cahill, Jr., G.F., 1970. Starvation in man. *New Engl. J. Med.* 282, 668–675.

Castellino, P., Solini, A., Luzi, L., Barr, J.G., Smith, D.J., Petrides, A., Giordano, M., Carroll, C., DeFronzo, R.A., 1992. Glucose and amino acid metabolism in chronic renal failure: effect of insulin and amino acids. *Am. J. Physiol.* 262, F168–F176.

Cooper, A.J.L., Plum, F., 1987. Biochemistry and physiology of brain ammonia. *Physiol. Rev.* 67, 440–519.

Dantzler, 1989. *Comparative Physiology of the Vertebrate Kidney.* Springer-Verlag, Heidelberg, Germany.

Darveau, C.A., Suarez, R.K., Andrews, R.D., Hochachka, P.W., 2002. Allometric cascade as a unifying principle of body mass effects on metabolism. *Nature* 417, 166–170.

Davis, P.K., Wu, G., 1998. Compartmentation and kinetics of urea cycle enzymes in procine enterocytes. *Comp. Biochem. Physiol. Part B* 119, 527–237.

Duchesne, R., Klein, J.D., Velotta, J.B., Doran, J.J., Rouillard, P., Roberts, B.R., McDonough, A.A., Sands, J.M., 2001. UT-A urea transporter protein in heart. Increased abundance during uremia, hypertension and heart failure. *Circ. Res.* 89, 139–145.

Frerichs, K.U., Smith, C.B., Brenner, M., DeGracia, D.J., Krause, G.S., Marrone, L., Dever, T.E., Hallenbeck, J.M., 1998. Suppression of protein synthesis in brain during hibernation involves inhibition of protein initiation and elongation. *Proc. Natl. Acad. Sci. USA* 95, 14511–14516.

Goodship, T.H.J., Mitch, W.E., Hoerr, R.A., Wagner, D.A., Steinman, T.I., Young, V.R., 1990. Adaptation to low-protein diets in renal failure: leucine turnover and nitrogen balance. *J. Am. Soc. Nephrol.* 1, 66–75.

Guarnieri, G., Antonione, R., Biolo, G., 2003. Mechanisms of malnutrition in uremia. *J. Renal Nutri.* 13, 153–157.

Harlow, H.J., Lohuis, T., Anderson-Sprecher, R.C., Beck, T.D.I., 2004. Body surface temperature of hibernating black bears may be related to periodic muscle activity. *J. Mammal.* 85, 414–419.

Heldmaier, G., Ortmann, S., Elvert, R., 2004. Natural hypometabolism during hibernation and daily torpor in mammals. *Resp. Physiol. Neurobiol.* 141, 317–329.

Hellgren, E.C., 1998. Physiology of hibernation in bears. *Ursus* 10, 467–477.

Hu, M-C., Bankir, L., Michelet, S., Rousselet, G., Trinh-Trang-Tan, M.-M., 2000. Massive reduction of urea transporters in remnant kidney and brain of uremic rats. *Kidney Int.* 58, 1202–1210.

Jackson, A.A., 1998. Salvage of urea-nitrogen in the large bowel: functional significance in metabolic control and adaptation. *Biochem. Soc. Trans.* 26, 231–236.

Jaroslow, B.N., Ortiz-Ortiz, I., Serrell, B.A., 1968. Protein synthesis during hibernation. *ANL Rep.* Dec, 102–103.

Klein, J.D., Timmer, R.T., Rouillard, P., Bailey, J.L., Sands, J.M., 1999. UT-A urea transporter protein expressed in liver: upregulation by uremia. *J. Am. Soc. Nephrol.* 10, 2076–2083.

Klein, J.D., Rouillard, P., Roberts, B.R., Sands, J.M., 2002. Acidosis mediates the upregulation of UT-A protein in livers from uremic rats. *J. Am. Soc. Nephrol.* 13, 581–587.

Koebel, D.A., Miers, P.G., Nelson, R.A., Steffen, J.M., 1991. Biochemical changes in skeletal muscles of denning bears (*Ursus americanus*). *Comp. Biochem. Physiol.* 100B, 377–380.

Kopple, J.D., Coburn, J.W., 1973. Metabolic studies of low protein diets in uremia. 1. Nitrogen and potassium. *Medicine* 52, 583–595.

Kopple, J.D., Greene, T., Chumlea, W.C., Hollinger, D., Maroni, B.J., Merrill, D., Scherch, L.K., Schulman, G., Wang, S-R., Zimmer, G.S., 2000. Relationship between nutritional status and the glomerular filtration rate: results from the MDRD study. *Kidney Int.* 57, 1688–1703.

Krapf, R., Seldin, D.W., Alpern, R.J., 2000. Clinical syndromes of metabolic acidosis. In: Seldin, D.W., Giebisch, G. (Eds.), *The Kidney. Physiology and Pathophysiology*, 3rd Edn. Lippincott Williams and Wilkins, Philadelphia, PA, pp. 2073–2130.

Kuhlmann, U., Schwickardi, M., Trebst, R., Lange, H., 2001. Resting metabolic rate in chronic renal failure. *J. Renal Nutri.* 11, 202–206.

Levinsky, N.G., Berliner, R.W., 1959. Changes in composition of the urine in ureter and bladder at low urine flow. *Am. J. Physiol.* 196, 549–553.

Li, J.B., Higgins, J.E., Jefferson, L.S., 1979. Changes in protein turnover in skeletal muscle in response to fasting. *Am. J. Physiol.* 236, E222–E228.

Lindholm, B., Heimburger, O., Stenvinkel, P., 2002. What are the causes of protein-energy malnutrition in chronic renal insufficiency? *Am. J. Kidney Dis.* 39, 422–425.

Liu, Z., Barrett, E.J., 2002. Human protein metabolism: its measurement and regulation. *Am. J. Physiol. Endocrinol. Metab.* 283, E1105–E1112.

Lundberg, D.A., Nelson, R.A., Wahner, H.W., Jones, J.D., 1976. Protein metabolism in the black bear before and during hibernation. *Mayo Clin. Proc.* 51, 716–722.

Marini, J.C., Van Amburgh, M.E., 2003. Nitrogen metabolism and recycling in Holstein heifers. *J. Anim. Sci.* 81, 545–552.

Marini, J.C., Klein, J.D., Sands, J.M., Van Amburgh, M.E., 2004. Effect of nitrogen intake on nitrogen recycling and urea transporter abundance in lambs. *J. Anim. Sci.* 82, 1157–1164.

Mayes, P.A., 2000. Digestion and absorption. In: Murray, R.K., Granner, D.K., Mayes, P.A., Rodwell, V.W. (Eds.), *Harper's Biochemistry*, 25th Edn. Appleton and Lange, Stamford, Connecticut, pp. 662–674.

McMahon, L.P., Parfrey, P.S., 2004. Cardiovascular aspects of chronic kidney disease. In: Brenner, B.M. (Eds.), *Brenner and Rector's the Kidney*, 7th Edn. Saunders, Philadelphia, PA, pp. 2189–2226.

Meredith, M., Layman, D.K., Baker, D.H., Nelson, R.A., 1988. Threonine metabolism in black bears. *FASEB J.* 2, A1530 (abstract #7166).

Metges, C.C., El-Khoury, A.E., Henneman, L., Petzke, K.J., Grant, I., Bedri, S., Pereira, P.P., Ajami, A.M., Fuller, M.F., Young, V.R., 1999, Availability of intestinal microbial lysine for whole body lysine homeostasis in human subjects. *Am. J. Physiol.* 277, E597–E607.

Metges, C.C., 2000. Contribution of microbial amino acids to amino acid homeostasis of the host. *J. Nutr.* 130, 1857S–1864S.

Monteon, F.J., Laidlaw, S.A., Shaib, J.K., Kopple, J.D., 1986. Energy expenditure in patients with chronic renal failure. *Kidney Int.* 30, 741–747.

Morrison, M., Mackie, R.I., 1997. Biosynthesis of nitrogen-containing compounds. In: Mackie, R.I., White, B.A. (Eds.), *Gastrointestinal Microbiology, Volume 1. Gastrointestinal Ecosystems and Fermentations*, Chapman and Hall, New York, pp. 424–469.

Motil, K.J., Matthews, D.E., Bier, D.M., Burke, J.F., Munro, H.N., Young, V.R., 1981. Whole-body leucine and lysine metabolism: response to dietary protein intake in young men. *Am. J. Physiol.* 240, E712–E721.

Nelson, R.A., Wahner, H.W., Jones, J.D., Ellefson, R.D., Zollman, P.E., 1973. Metabolism of bears before, during, and after winter sleep. *Am. J. Physiol.* 224, 491–496.

Nelson, R.A., Jones, J.D., Wahner, H.W., McGill, D.B., Code, C.F., 1975. Nitrogen metabolism in bears: urea metabolism in summer starvation and in winter sleep and role of urinary bladder in water and nitrogen conservation. *Mayo Clin. Proc.* 50, 141–146.

Nelson, R.A., Beck, T.D.I., Steiger, D.L., 1984. Ratio of serum urea to serum creatinine in wild black bears. *Science* 226, 841–842.

Nelson, R.A., Anderson, C.F., Hunt, J.C., Margie, J., 1985. Nutritional management of chronic renal failure for two purposes: postponing onset and reducing frequency of dialysis. In: Cummings, N.B., Klahr, S. (Eds.), *Chronic Renal Disease*, Plenum, New York, pp. 573–585.

Nelson, R.A., 1989. Nitrogen turnover and its conservation in hibernation. In: Malan, A., Canguilhem, B. (Eds.), *Living in the Cold*. John Libbey Eurotext Ltd, Montrouge, France, pp. 299–307.

O'Sullivan, A.J., Lawson, J.A., Chan, M., Kelly, J.J., 2002. Body composition and energy metabolism in chronic renal insufficiency. *Am. J. Kidney Dis.* 39, 369–375.

Owen, O.E., Felig, P., Morgan, A.P., Wahren, J., Cahill Jr., G.F., 1969. Liver and kidney metabolism during prolonged starvation. *J. Clin. Invest.* 48, 574–583.

Panesar, A., Agarwal, R., 2003. Resting energy expenditure in chronic kidney disease: relationship with glomerular filtration rate. *Clin. Nephrol.* 59, 360–366.

Reaich, D., Channon, S.M., Scrimgeour, C.M., Goodship, T.H.J., 1992. Ammonium chloride-induced acidosis increases protein breakdown and amino acid oxidation in humans. *Am. J. Physiol.* 263, E735–E739.

Reaich, D., Channon, S.M., Scrimgeour, C.M., Daley, S.E., Wilkinson, R., Goodship, T.H.J., 1993. Correction of acidosis in humans with CRF decreases protein degradation and amino acid oxidation. *Am. J. Physiol.* 265, E230–E235.

Rodwell, V.W., 2000a. Biosynthesis of the nutritionally non-essential amino acids. In: Murray, R.K., Granner, D.K., Mayes, P.A., Rodwell, V.W. (Eds.), *Harper's Biochemistry*, 25th Edn. Appleton and Lange, Stamford, Connecticut, pp. 307–312.

Rodwell, V.W., 2000b. Catabolism of proteins and of amino acid nitrogen. In: Murray, R.K., Granner, D.K. Mayes, P.A., Rodwell, V.W. (Eds.),

Harper's Biochemistry, 25th Edn. Appleton and Lange, Stamford, Connecticut, pp. 313–322.

Rolfe, D.F.S., Brown, G.C., 1997. Cellular energy utilization and molecular origin of standard metabolic rate in mammals. *Physiol. Rev.* 77, 731–758.

Sands, J.M., 1999. Regulation of renal urea transporters. *J. Am. Soc. Nephrol.* 10, 635–646.

Sands, J.M., 2002. Molecular approaches to urea transporters. *J. Am. Soc. Nephrol.* 13, 2795–2806.

Sands, J.M., 2003. Mammalian urea transporters. *Annu. Rev. Physiol.* 65, 543–566.

Schneeweiss, B., Graninger, W., Stockenhuber, F., Druml, W., Ferenci, P., Eichinger, S., Grimm, G., Laggner, A.N., Lenz, K., 1990. Energy metabolism in acute and chronic renal failure. *Am. J. Clin. Nutri.* 52, 596–601.

Schulman, G., Himmelfarb, J., 2004. Hemodialysis. In: Brenner, B.M. (Ed.), *Brenner and Rector's The Kidney*, 7th Edn. Saunders, Philadelphia, PA, pp. 2563–2624.

Singer, M.A., 2001a. Of mice and men and elephants: metabolic rate sets glomerular filtration rate. *Am. J. Kidney Dis.* 37, 164–178.

Singer, M.A., 2001b. Ammonia functions as a regulatory molecule to mediate adjustments in glomerular filtration rate in response to changes in metabolic rate. *Med. Hypotheses* 57, 740–744.

Singer, M.A., 2002. Vampire bat, shrew, and bear: comparative physiology and chronic renal failure. *Am. J. Physiol. Regul. Integ. Comp. Physiol.* 282, R1583–R1592.

Singer, M.A., 2003a. Dietary protein-induced changes in excretory function: a general animal design feature. *Comp. Biochem. Physiol. Part B* 136, 785–801.

Singer, M.A., 2003b. Do mammals, birds, reptiles and fish have similar nitrogen conserving systems? *Comp. Biochem. Physiol. Part B* 134, 543–558.

Smith, C.P., Potter, E.A., Fenton, R.A., Stewart, G.S., 2004. Characterization of a human colonic cDNA encoding a structurally novel urea transporter, hUT-A6. *Am. J. Physiol. Cell Physiol.* 287, C1087–C1093.

Stevens, C.E., Hume, I.D., 1998. Contributions of microbes in vertebrate gastrointestinal tract to production and conservation of nutrients. *Physiol. Rev.* 78, 393–427.

Stewart, G.S., Fenton, R.A., Thevenod, F., Smith, C.P., 2004. Urea movement across mouse colonic plasma membranes is mediated by UT-A urea transporters. *Gastroenterology* 126, 765–773.

Storey, K.B., 2003. Mammalian hibernation, transcriptional and translational controls. *Adv. Exp. Med. Biol.* 543, 21–38.

Taal, M.W., Luyckx, V.A., Brenner, B.M., 2004. Adaptation to nephron loss. In: Brenner, B.M. (Ed.), *Brenner and Rector's The Kidney*, 7th Edn. Saunders, Philadelphia, PA, pp. 1955–1997.

Terris, J., Ecelbarger, C.A., Sands, J.M., Knepper, M.A., 1998. Long-term regulation of renal urea transporter protein expression in rat. *J. Am. Soc. Nephrol.* 9, 729–736.

Tinker, D.B., Harlow, H.J., Beck, T.D.I., 1998. Protein use and muscle-fiber changes in free-ranging, hibernating black bears. *Physiol. Zool.* 71, 414–424.

Torrallardona, D., Harris, C.I., Coates, M.E., Fuller, M.F., 1996. Microbial amino acid synthesis and utilization in rats: incorporation of ^{15}N from ^{15}NH$_4$Cl into lysine in the tissues of germ-free and conventional rats. *Br. J. Nutri.* 76, 689–700.

Torrallardona, D., Harris, C.I., Fuller, M.F., 2003a. Pigs' gastrointestinal microflora provide them with essential amino acids. *J. Nutri.* 133, 1127–1131.

Torrallardona, D., Harris, C.I., Fuller, M.F., 2003b. Lysine synthesized by the gastrointestinal microflora of pigs is absorbed, mostly in the small intestine. *Am. J. Physiol. Endocrinol. Metab.* 284, E1177–E1180.

Varcoe, R., Halliday, D., Carson, E.R., Richards, P., Tavill, A.S., 1975. Efficiency of utilization of urea nitrogen for albumin synthesis by chronically uraemic and normal man. *Clin. Sci. Molec. Med.* 48, 379–390.

Virtanen, A.I., 1966. Milk production of cows on protein-free feed. *Science* 153, 1603–1614.

Walser, M., Bodenlos, L.J., 1959. Urea metabolism in man. *J. Clin. Invest.* 38, 1617–1626.

Walser, M., 1974. Urea metabolism in chronic renal failure. *J. Clin. Invest.* 53, 1385–1392.

Waterlow, J.C., 1984. Protein turnover with special reference to man. *Quart. J. Exper. Physiol.* 69, 409–438.

Waterlow, J.C., 1999. The mysteries of nitrogen balance. *Nutri. Res. Rev.* 12, 25–54.

Wolfe, R.R., Nelson, R.A., Stein, T.P., Rogers, L., Wolfe, M.H., 1982. Urea nitrogen reutilization in hibernating bears. *Fed. Proc.* 41, 1623 (abstract #7908).

Wright, P.A., Obbard, M.E., Battersby, B.J., Felskie, A.K., LeBlanc, P.J., Ballantyne, J.S., 1999. Lactation during hibernation in wild black bears: effects on plasma amino acids and nitrogen metabolites. *Physiol. Biochem. Zool.* 72, 597–604.

Wu, G., 1995. Urea synthesis in enterocytes of developing pigs. *Biochem. J.* 312, 717–723.

Yacoe, M.E., 1983. Protein metabolism in the pectoralis muscle and liver of hibernating bats, *Eptesicus fuscus. J. Comp. Physiol.* 152, 137–144.

Young, V.R., Haverberg, L.N., Bilmazes, C., Munro, H.N., 1973. Potential use of 3-methylhistidine excretion as an index of progressive reduction in muscle protein catabolism during starvation. *Metabolism* 22, 1429–1436.

Young, V.R., El-Khoury, A.E., 1994. Is leucine produced by the colonic microflora? Reply to A.A. Jackson. *Am. J. Clin. Nutri.* 60, 978–979.

CHAPTER 3

Atherosclerotic Vascular Disease

Introduction

Atherosclerosis is a disease of large and medium sized elastic and muscular arteries (Ross, 1999). It has been estimated that this disease is responsible for over 55% of all deaths in Western civilization (Owens *et al.*, 2004) primarily due to ischemic damage of end organs such as the heart, brain and kidney. The advanced complicated atherosclerotic lesion consists of a fibrous cap (which projects into the arterial lumen) with embedded smooth muscle cells overlying a core of lipid, macrophages and necrotic debris (Ross, 1993). As pointed out by Ross (1999), the lesions of atherosclerosis represent a series of highly specific cellular and molecular responses that can best be described, in aggregate, as an inflammatory disease. The natural animal models considered in this chapter are the salmon and trout. These bony fish (teleosts) develop arteriosclerotic lesions in the coronary artery, which appear to resemble histologically an early stage of mammalian atherosclerosis. However, the salmon/trout lesions do not display an inflammatory component and do not progress to the stage of advanced complicated atherosclerotic lesions as observed in the mammal. Hence, in this chapter, we will consider the following question: are the coronary arteriosclerotic lesions of the salmon and trout a realistic model in terms of genesis and morphology of the early atherosclerotic lesions of mammals, and if so why do these lesions not develop into the advanced complicated lesions found in mammals? This question complements that generally addressed with experimental animal models of atherosclerosis, which is: what are the factors promoting progression of the lesions. In essence, a histological change (lesion), which is clearly adaptive and non-progressive in the salmon and trout, becomes a progressive disease process in mammals resulting in significant tissue damage.

Atherosclerosis in Mammals

Normal arteries are composed of three morphologically distinct layers (Ross and Glomset, 1973). The intima (inner layer) is bounded on one side by a single layer of endothelial cells that line the lumen and on the other side by the internal elastic lamina. Between these two boundaries are smooth muscle cells and extracellular matrix. The media (middle layer) consists of a large number of smooth muscle cells each surrounded by small amounts of collagen and varying numbers of small elastic fibers and other extracellular matrix components. The outer layer or adventitia consists mainly of fibroblasts, collagen and mucopolysaccharides. The adventitia is separated from the media by a poorly defined sheet of elastic tissue — the external elastic lamina. Fibroblasts are not found in the intima or media of mammalian arteries. Ross and Glomset in 1973 described the probable histological changes in the development of an atherosclerotic lesion. According to this model, the first phase consists of a focal thickening of the intima due to an increase in smooth muscle cells and extracellular matrix. Smooth muscle cells proliferate within the intima and also migrate from the media into the intima. Lipids accumulate both within smooth muscle cells and between cells. Further development of the lesion involves formation of a fibrous cap and ingress of blood constituents. According to Ross and Glomset (1973) smooth muscle cells play a fundamental role in the genesis of atherosclerotic lesions. They further postulated that the initiating event was endothelial cell injury with a decrease in the normal barrier properties of this cell layer. A decrease in the endothelial barrier would allow blood constituents to enter the intima. As part of the injury response smooth muscle cells migrate from the media into the intima and proliferate. If the endothelial injury is self limiting the lesion can regress. If the endothelial injury persists then smooth muscle cells continue to proliferate with further development of the atherosclerotic lesion.

The hypothesis that an injury to the endothelium precipitates the atherosclerotic process has been considerably modified since the paper by Ross and Glomset in 1973. The initial injurious events do not necessarily result in endothelial denudation and in fact early lesions develop at sites of morphologically intact endothelium (Ross, 1993). A recent version of this hypothesis emphasizes endothelial dysfunction rather than denudation. Endothelial cells play numerous physiological roles such as: provision of a non-thrombogenic surface, a permeability barrier, maintenance of vascular tone, formation and secretion of growth regulating molecules and cytokines, and provision of a non-adherent surface for leukocytes. Changes in one or more of these properties may represent the earliest manifestations of endothelial dysfunction (Ross, 1993). Possible causes of endothelial dysfunction (leading to atherosclerosis) include physical forces such as alterations in blood flow, elevated and modified low density lipoproteins (LDL), free radicals caused by cigarette smoke, hypertension, diabetes mellitus, genetic alterations, elevated plasma homocysteine concentrations and infectious microorganisms (Ross, 1999). As a consequence of alterations in the properties of the endothelial layer, a series of responses occur which impair the normal homeostatic functions of the endothelium. These alterations include: increased endothelial permeability, upregulation of various leukocyte adhesion molecules and migration of leukocytes into the arterial wall (Ross, 1999). Three cellular components of the circulation (monocytes, platelets and T lymphocytes) together with two cell types of the arterial wall (endothelial and smooth muscle cells) interact in multiple ways in generating atherosclerotic lesions (Osterud and Bjorklid, 2003).

The current model for the pathogenesis of atherosclerotic lesions has been discussed in a number of recent review articles (Doherty *et al.*, 2003; Lusis, 2000; Ross, 1993; Ross, 1999; Libby, 2002; Osterud and Bjorklid, 2003; Ohashi *et al.*, 2004). The following brief description is an amalgam of these articles.

Specific arterial sites, such as branches, bifurcations and curvatures cause characteristic alterations in the flow of blood with changes in wall shear stresses and these alterations probably account for the location of initial atherosclerotic lesions. Within this context a working hypothesis has been advanced that the endothelial lining is the primary sensor of wall shear stresses and functions as a transducer of these biomechanical stimuli into biological responses within the vessel wall (Gimbrone, 1999). These biological responses (collectively referred to as endothelial dysfunction) include production of cytokines, growth factors and extracellular matrix components. Systemic risk factors linked to endothelial dysfunction have already been listed, but LDL is considered to play the dominant role in the genesis of lesions perhaps because the current model is based mainly on observations in experimental animals with a high rate of developing lesions, i.e. fat-fed and genetically hyperlipidemic. The primary initiating event is the accumulation of LDL in the subendothelial matrix. Accumulation is greater when levels of circulating LDL are raised. The transport and retention of LDL are increased in the preferred sites for lesion formation. As already discussed these sites are determined by hemodynamic factors. It is believed that the trapped LDL is subsequently modified in the vessel wall by processes such as oxidation, glycation and aggregation. Hemodynamic factors plus trapped and modified LDL particles in the intima "activate" the endothelium. As a result the endothelial cells produce a number of pro-inflammatory molecules including leukocyte adhesion molecules, chemotactic proteins and growth factors. Monocytes and lymphocytes are recruited to the vessel wall. Once resident in the arterial intima, monocytes acquire the morphological characteristics of macrophages. LDL particles trapped in the intima are prone to progressive oxidation. This modification of LDL particles renders them recognizable by macrophage scavenger receptors and thus targets for internalization by these cells. Upon extensive uptake of modified LDL via scavenger receptors, macrophages are ultimately turned into foam cells and form macroscopically so-called fatty streaks. T lymphocytes also enter the intima facilitated by binding to adhesion molecules. T lymphocytes may also be involved in the formation of fatty streaks. In addition, macrophages and lymphocytes produce a number of cytokines and growth factors, which influence the behavior of other cells, present in the atheromatous lesion thereby leading to the development of more advanced lesions. Progression results in the formation of a fibrous plaque which is characterized by a growing mass of extracellular lipid (due to the death of foam cells) and by the accumulation of smooth muscle cells (SMC) and SMC derived extracellular matrix. Cytokines and growth factors secreted by macrophages and lymphocytes are believed to be important mediators of SMC migration, proliferation and extracellular matrix production. In addition, SMCs take up oxidized LDL to form lipid laden foam cells. Hence, the fatty streak is actually composed of foam cells derived from both SMCs and macrophages.

Several risk factors seem to contribute to the development of fibrous lesions. Elevated levels of homocysteine appear to injure endothelial cells and to stimulate proliferation of SMCs. Some of the effects of hypertension are mediated by components of the renin angiotensin system. For example, angiotensin II directly stimulates SMC growth and the production of extracellular matrix.

Platelets have also been implicated in the genesis of atherosclerotic lesions. Platelets can adhere to dysfunctional endothelium and when activated release a number of cytokines and growth factors (as do macrophages and lymphocytes). For example, platelet derived growth factor stimulates SMC proliferation and migration. In addition, platelets can instigate thrombus formation when exposed to the subendothelial intimal compartment.

In summary, the current model places a heavy emphasis on the role of modified LDL particles trapped in the subendothelial space. These particles are pivotal in activating endothelial cells, which then secrete a number of pro-inflammatory molecules. Monocytes and lymphocytes are recruited to the site. An inflammatory reaction proceeds with subsequent involvement of platelets and SMCs. This model is supported for example, by studies in mice lacking Apolipoprotein E (Zhang *et al.*, 1992). Homozygous mutant apoE deficient mice show elevated total plasma cholesterol levels about five times that of normal mice with a reduced concentration of high density lipoproteins (HDL). Nearly 80% of the total cholesterol is carried in particles having sizes that are similar to the lower density lipoproteins of humans. These mice spontaneously develop vascular lesions similar to those observed during atherogenesis in humans (Nakashima *et al.*, 1994; Zhang *et al.*, 1992). Early changes consisted of adherence of mononuclear cells to the endothelial surface. Fatty streaks consisting of lipid-laden monocyte-derived macrophages subsequently formed. With progression, SMCs appeared, many of which contained lipid deposits. Ultimately, fibrous caps rich in SMCs formed over the foam cell rich areas. With further progression, cholesterol clefts and necrotic areas appeared.

However, there are a number of existing observations which are not completely consistent with this current model, particularly the early initiating events in the genesis of lesions. In order to place these observations in context, we need to review some relevant aspects of the biology of the arterial tree.

Biology of the Arterial System: Implications for the Genesis of Atherosclerosis in Mammals

As will be discussed in the following section, there are observations consistent with the notion that pre-atherosclerotic lesions occur *in utero*. Hence, we must review some of the factors regulating arterial vessel development in the embryo. As pointed out by Burggren (2004), dogma holds that the embryonic heart begins to beat in order to pump blood for convective transport of oxygen, nutrients and metabolic wastes as it does in the adult. However, there is compelling evidence that part of the initial rationale for the beat of the vertebrate embryonic heart is to aid in cardiac morphogenesis and to promote the formation of new vessels by sprouting from vessel tips. Burggren (2004) proposed that early pulsatile flow creates shear/strain on endothelial cells lining sprouting arterial blood vessels. In response endothelial cells proliferate under the influence of vascular endothelial growth factor and other secreted factors, resulting in the branching and sprouting of new vessels. Hence according to this hypothesis, pressure/flow relationships are important determinants of arterial blood vessel network development. Abnormalities of pressure/flow relationships *in utero* could trigger a variety of metabolic responses by the endothelial cells. As already discussed in the "Atherosclerosis in Mammals" section, the metabolic responses of endothelial cells to biomechanical stimuli are probably an important event in the genesis of atherosclerotic lesions (Gimbrone, 1999). Although biomechanical factors are an important determinant of vascular development, in the coronary vascular bed, for example, much of the initial differentiation and patterning of this system occurs in the absence of blood flow (Wada *et al.*, 2003). Also the first steps in the diversification of coronary vessels into arteries, veins and capillaries occur before initiation of blood flow. Thus in the coronary system, vasculogenesis is driven by a program of differentiation intrinsic to the developing coronary vessels. At the end of the vasculogenic period, without blood flow, the general pattern of the coronary system is set. However, significant remodeling of the major vessels and capillaries does take place after connection to the aorta and the initiation of blood flow.

Since the intima is the soil for the early atherosclerotic lesion, we must review specifically the biology of this layer of the arterial wall. The intima is defined as the region of the arterial wall from and including the endothelial surface at the lumen to the luminal margin of the media (Stary *et al.*, 1992). The internal elastic lamina denotes the border between intima and media although a well-defined internal elastic lamina may not be present at bifurcations, branch vessels and curvatures. According to Stary *et al.* (1992) the intima contains two layers: an inner layer below the endothelium composed of proteoglycan ground substance and a scattering of smooth muscle cells (SMC) and an outer layer containing abundant SMCs and elastic fibers (musculoelastic layer). However this description of the intima is somewhat controversial. According to Ikari *et al.* (1999) the musculoelastic layer actually belongs to the media and not the intima. The thickness of the intima is not uniform and both focal and diffuse patterns of thickening have been described. The significance of intimal thickening, especially the focal pattern is discussed below. The histological features of intimal thickening are somewhat variable and age related but a description has been given by Stary (1989). Focal intimal thickening (sometimes referred to as intimal cushions or pads) consists of an upper layer of increased proteoglycan rich matrix containing sparse elastic fibers and scattered SMCs and a much-thickened lower musculoelastic layer containing compact rows of SMCs and elastic fibers. Focal thickenings generally occur at arterial bifurcations and vessel orifices. Diffuse intimal thickening is often circumferential and not clearly related to specific geometrical configurations of arteries. Histologically, diffuse intimal thickenings contain the same components as focal thickenings but of proportionately lesser amounts (Stary, 1989). This description of the arterial intima applies generally to large mammals (French, 1966). In small mammals, the arterial intima has a predominately simple structure throughout life. For example, in the mouse and rat, the endothelium of the aorta appears to be in contact with the internal elastic lamina (French, 1966).

Several investigators have addressed the relationship between intimal thickening, specifically focal thickening occurring in the fetal and neonatal periods and the subsequent development of atherosclerotic lesions. These studies have in general been limited to two atherosclerotic prone sites — the coronary and internal carotid arteries.

Ikari *et al.* (1999) studied intima formation in the human coronary system. They examined the proximal left anterior descending coronary artery but specifically excluded bifurcations since intimal cushions start at bifurcations. Intima was detected at three months before birth in about 15% of specimens. Intima was found in 38% of specimens just after birth and in all specimens at three months of age. The intima of these young coronary arteries consisted of smooth muscle cells — no macrophages were detected. At bifurcations in the coronary artery tree, focal intimal thickening can be found in four-month-old fetuses (Velican and Velican, 1976).These focal thickenings or cushions develop prior to true intima formation in the straight running portions of the coronary vasculature. In these fetuses, the wall of the left coronary trunk contained endothelial cells lying directly upon the internal elastic membrane. However, at the emergence zone of the anterior descending branch of the left coronary artery focal intimal thickenings were present. These thickenings or cushions were characterized by splitting of the internal elastic membrane and accumulation of medial SMCs. These cushions changed in an age related manner. In the six-month-old fetus, these cushions acquired a more inhomogeneous structure with numerous longitudinally arranged bundles of SMCs in their basal segment. At birth, the degree of arterial thickening and extension from the emergence area of the anterior descending branch of the left coronary artery was considerably advanced compared to other vessels of the coronary tree and compared to the basilar and renal arteries belonging to the same subject. Although not specifically stated, the implication given in this paper is that all of the fetuses examined showed these focal areas of intimal thickening. This study by Velican and Velican (1976) also included observations on 50 children ages one to ten. Approximately one third showed the development of a mucoid connective tissue with a honeycomb structure in the subendothelial zone of the thickened intima. This change only appeared in the thickened intima of the proximal segment of the anterior descending branch of the left coronary artery. In 16% of the children, a fibrotic intima was apparent in the emergence region of the anterior descending branch. Eight percent of children had small areas of necrosis in the deep intimal regions of the proximal segment of the anterior descending branch. Two children also exhibited very small lipid droplets, in the respective areas, intermingled with the necrotic debris. No similar pathologic changes were detected in other coronary vessels or in the basilar and renal arteries of the same subject. In summary the studies of Velican and Velican (1976), demonstrate that in the coronary arteries, focal intimal thickenings develop *in utero* in a non-random distribution (at branch points) and predominantly in the left coronary artery system. These thickenings show age related complex histological changes with fibrosis and small areas of necrosis present in children ages one to ten years. The concordance of the distribution of these neonatal intimal masses with the distribution of atherosclerotic lesions is striking. As stated by the authors: "Our results show that the particular susceptibility of the artery of sudden death (anterior descending branch of the left coronary artery) to atherosclerotic involvement has a counterpart in the precocity, intensity and severity of the early age-dependent changes detected in the prenatal and postnatal periods investigated."

In a later study, Velican and Velican (1979) described the microarchitecture of human coronary arteries and included a comparative study of the basilar, anterior cerebral, renal, hepatic and bronchial arteries. As pointed out by these authors, blood flow in the coronary arteries differs from that in other organ arteries in that it is intermittent. During systole the contracting heart compresses intramyocardial vessels and impedes blood flow. Also coronary vessels are unique in that they are rhythmically subjected to torsion, spiral twisting and bending with every myocardial contraction. Hence, the hemodynamic/biomechanical forces associated with the coronary arteries differ from those of other organ arteries. Consistent with this difference, an intimal layer develops in the coronary artery during fetal life (as already discussed) whereas in other organ arteries an intimal layer can be seen only in adolescents (basilar and renal arteries) or in young adults (anterior cerebral, hepatic and bronchial arteries). In the coronary arteries, large intimal thickenings or cushions developed at branch points in fetal life and after birth they coalesced giving rise to a diffuse thickened intima at bifurcations and emergence areas. In addition to a very thick intima and large cushions, the branching areas of human coronary arteries exhibited a particular development and microarchitecture of longitudinal muscle columns not seen in other organ arteries. In the basilar, anterior cerebral, renal, hepatic and bronchial arteries intimal branch pads or cushions occurred only in postnatal life. During infancy, childhood and adolescence these cushions appeared circumscribed at arterial forks and with age exhibited only a moderate increase in thickness.

A similar study has been conducted by Stary (1989). He studied the left coronary artery (proximal part including the take off sections of the circumflex and left anterior descending branches) in 565 subjects who died between full term and age 29 years. Focal intimal thickenings were restricted to bifurcations and were present from birth. Stary reported age related changes in the histology of the coronary artery intima, which were, either much more frequent or restricted to sites of focal intimal thickening. In 45% of infants (first eight months of life) macrophage foam cells were detected. These cells were five times more numerous in areas of focal thickening and were present deep in the subendothelial space. About eight percent of children between ages 12 and 14 years had preatheroma lesions consisting of lipid-laden macrophages and SMCs as well as extracellular lipid particles. Preatheroma were limited to coronary artery locations with focal thickenings. Atheromas were observed in somewhat older subjects. These lesions contained all the components of preatheroma but in addition the extracellular lipid particles were fused to form a massive lipid core. Atheromas were restricted to sites of focal thickening. Preatheroma, atheroma or fibroatheroma were observed in about one third of subjects aged 27 to 29 years.

Studies in a second atherosclerotic prone site, the internal carotid system, complement the observations described in the coronary arteries. The parasellar internal carotid artery (pICA) has a high propensity toward atherosclerosis and lesions in this location frequently cause a stroke. Weninger *et al.* (1999), studied the pICA in human infants ages three weeks to nine months. Focal intimal thickenings or cushions occurred at three characteristic locations; two of which were curves in the course of the pICA. The cushions contained matrix, elastic fiber bundles and a variable number of SMCs. In some circumscribed areas, the internal elastic lamina was ruptured or split at the fringes. In non-hyperplastic unaffected regions of the pICA, the intima consisted of a single endothelial cell layer separated from the media by an internal elastic lamina. Intimal cushions were found in all specimens although the frequencies varied for the different locations. The adult pICA is a site highly prone to atherosclerosis and the occurrence of intimal masses in the infant pICA clearly suggests that that these two phenomena are linked. Although hemodynamic factors are thought to be the major trigger for the genesis of focal intimal thickening (Stehbens, 1996), Weninger *et al.* (1999), propose that in the pICA, intimal cushions are associated with developmental shape transformations of this vessel rather than primarily with blood flow patterns. They found a statistically significant correlation between the shape of the pICA and the degree of intimal hyperplasia.

The relationship between intimal masses and atherosclerotic lesions was examined experimentally by Scott *et al.* (1979) in young swine. When atherosclerosis was induced in the aortas of young swine fed a hypercholesterolemic diet, the distribution of early lesions was similar to the distribution of intimal cell masses in control mash-fed swine. The investigators concluded that the early atherosclerotic lesions were derived from previously existing intimal cell masses.

All of these observations strongly suggest that the intima or more specifically the sites of focal thickening provide a receptive "soil" for the genesis of atherosclerotic lesions although the particular properties of the intima that account for this receptive environment have not been identified (Schwartz *et al.*, 1995; Schwartz, 1999). A simple working hypothesis is that the cells making up these masses have special (as yet unknown) properties. The most important cells making up intimal masses are SMCs and there is now evidence that intimal SMCs are a distinctly different cell type from the medial SMC (Schwartz *et al.*, 1995; Mulvihill *et al.*, 2004).

The complexity of the early events in the genesis of the atherosclerotic lesion is compounded by the dynamic properties of SMCs. As reviewed by Owens *et al.* (2004), SMCs within adult animals retain a remarkable plasticity and can undergo rather profound and reversible changes in phenotype in response to changes in local environmental cues. One of these local cues involves a regulatory function of elastin (Li *et al.*, 1998). Mice lacking elastin develop obstructive arterial disease resulting from subendothelial SMC proliferation and reorganization. Lack of elastin was not associated with obvious endothelial damage, thrombosis or inflammation. Hemodynamic factors were not responsible for the SMC proliferation since SMC proliferation still occurred in aortic segments (from mice lacking elastin) isolated in organ culture and therefore not subject to hemodynamic stress.

The morphological, biochemical, physiological and molecular properties of the SMC vary at different stages of atherosclerosis (Owens *et al.*, 2004). The source of intimal SMC has not been completely established. The best-studied model of neointimal formation is the response of the rat carotid artery to balloon angioplasty. In this model, medial SMCs proliferate and migrate across the internal elastic lamina to form the intima (Schwartz *et al.*, 1995). However this model may not accurately reflect events underlying true intima formation. As already indicated, there is evidence that intimal SMCs are probably a distinct cell type and not derived from medial SMCs.

Genesis of Atherosclerotic Lesions

I believe the evidence is compelling that the intimal cushions described at branch points, vessel orifices and curvatures in the neonatal coronary and internal carotid arteries are related to the subsequent development of atherosclerotic lesions. These focal thickenings represent a response to hemodynamic forces and/or shape alterations in the developing vessel. These thickenings occur, at least in the coronary vessels, prior to the formation of "true" intima in the straight running portions of the arteries. These cushions demonstrate age-related histological changes with atheroma developing, according to the study of Stary (1989), in up to one third of subjects in their late twenties. The factors regulating this intimal response are largely unknown, although elastin itself appears to have a regulatory role in this process. These intimal cushions are composed of matrix and SMCs. Initial studies implied that these SMCs were probably medial SMCs that had proliferated and migrated into the intima. However more recent evidence suggests that the intimal SMC is a distinct cell type from the medial SMC. Hence, events within the arterial intima in the perinatal period somehow determine the subsequent non-random localization of atherosclerotic lesions in adults. These early perinatal intimal cushions provide a soil susceptible to the atherosclerotic process. However, neither the nature of this susceptibility or the mechanism underlying the long latent period between formation of intimal cushions and development of atherosclerotic lesions are known. Furthermore, these observations with respect to the role of focal intimal thickenings or cushions are not explicitly included in the current model for the pathogenesis of atherosclerosis described in the "Atherosclerosis in Mammals" section. However, could the study of a natural animal model unravel more of the details with respect to the process of atherogenesis?

Natural Animal Model: The Fish

As noted in the "Introduction" section, the natural animal model for atherosclerosis is the fish, particularly the salmon and trout but also the dogfish. These fish develop focal intimal thickenings in their coronary arteries which histologically are characterized by intimal proliferation of vascular smooth muscle cells with a disrupted internal elastic membrane (Farrell, 2002). Calcium and lipid deposits are not found in these lesions. These intimal thickenings in the fish coronary arteries have a striking resemblance to the focal intimal thickenings found in the coronary and internal carotid arteries of humans as described in the "Biology of the Arterial System: Implications for the Genesis of Atherosclerosis" section. As already discussed, these intimal cushions in the human provide the soil for the later development of atherosclerotic lesions. Hence, the histological similarity between human and fish focal intimal thickenings makes the latter a wonderful natural model for the earliest events in the genesis of atherosclerosis in humans. In the next sections, selected aspects of the fish coronary circulation as well as the histology and biology of coronary arteriosclerosis in fish are reviewed.

Fish Coronary Circulation

All elasmobranchs and about one third of teleost species have a coronary circulation (Franklin and Axelsson, 1994). In fact, elasmobranchs were the first vertebrates to evolve a coronary circulation. The coronary circulation arises post-branchially (i.e. after the gills) as cranial hypobranchial arteries coming from one to three pairs of the efferent branchial arteries. The hypobranchial arteries run along the ventral aorta and conus/bulbus arteriosus before ramifying through the myocardium. A key difference between the fish and the mammal is that the coronary system of the mammal arises from the base of the aorta and thus immediately and directly perfuses the myocardium, whereas in the fish the blood has to pass through the gill capillary bed where it is oxygenated before perfusing the coronary vessels.

Cyclostomes and most sedentary fish have a heart, which consists of spongy myocardial tissue only (Davie and Farrell, 1991). These fish have no coronary circulation and myocardial oxygenation depends upon diffusion of oxygen from blood within the cardiac chambers. Elasmobranchs and more active teleost species (about one third) do have a coronary circulation which functions to provide a supplementary supply of oxygen to cardiac muscle (Farrell, 2002). The myocardium in these fish consists of both an outer compact layer as well as an inner spongy layer. The coronary circulation has two basic patterns. In the salmon and trout for example, the coronary system supplies the compact layer but does not penetrate the inner spongy layer. Compact myocardium accounts for about 30% of ventricular mass in these fish (Farrell, 2002; Franklin and Axelsson, 1994). Elasmobranchs have a coronary circulation in both the compact and spongy myocardium (Franklin and Axelsson, 1994). Although the coronary circulation provides a supplemental supply of oxygen to cardiac muscle in addition to that supplied by diffusion from the cardiac lumen, several observations attest to the functional importance of the coronary system. Acute ligation of the main stem coronary artery in Chinook salmon (*Oncorhynshus tshawytscha*) results in a 35.5% reduction in maximum sustained swimming speed (Farrell and Steffensen, 1987). Hence, the fish coronary circulation is necessary for maximum aerobic swimming performance. In addition male rainbow trout develop ventricular hypertrophy as they reproductively mature (Clark et al.; 2004). This hypertrophy is accompanied by compensatory angiogenesis in the epicardial (coronary) capillaries as one mechanism for maintaining oxygenation in the hypertrophied ventricle.

Fish Coronary Artery Thickenings

In the literature, focal intimal thickenings in the fish are generally referred to as lesions. However, as will be discussed, they never develop the histological features of advanced atherosclerotic changes as observed in humans and these intimal thickenings are probably not pathologic. However, since usage of the term lesion is so widespread in the literature, the term lesion is also used in this chapter when referring to focal arterial intimal thickenings in the fish.

Focal intimal thickenings have been most thoroughly described in the coronary arteries of salmon and trout and to some extent elasmobranchs. However these lesions are not exclusive to the coronary arteries. For example, Robertson *et al.* (1961) described large nodules of intimal hyperplasia with destruction of the underlying internal elastic membrane in the arteries of the pancreas of spawning male Steelhead trout. The normal intima of the fish coronary artery consists of a layer of endothelial cells resting on the internal elastic lamina (Moore *et al.*, 1976a). Histologically the fish coronary artery intima resembles that of small mammals (French, 1966).

Maneche *et al.* (1972) reported on coronary artery lesions in Atlantic salmon (*Salmo salar*). The natural life cycle of this fish includes an initial developmental stage in fresh water, followed by a maturation period in the sea and culminates in the spawning by the mature fish after its return to fresh water. The Atlantic salmon is unique in that multiple spawning cycles and seaward migrations can take place during the natural life cycle. Lesions consisted of focal intimal proliferation of smooth muscle cells often associated with morphological alterations of the underlying internal elastic lamina which appeared thickened, fragmented or split in some areas. No fatty deposits were observed. Lesion severity was classified according to the degree of intimal proliferation and luminal narrowing. There was a high incidence of coronary artery lesions in both pre-spawners and post-spawners. However both the incidence and severity of lesions were greater in pre- as opposed to post-spawners. Maneche *et al.* (1972) concluded that considerable regression of lesions occurred after spawning implying that lesion formation was reversible.

Van Citters and Watson (1968) studied the coronary arteries of Steelhead trout, a species like the Atlantic salmon in which most survive spawning. Many return to the sea and some spawn a second or even a third time. Coronary lesions were absent in juvenile fish, which had not migrated downstream into salt water and were uncommon in immature fish at sea.

However the incidence and severity of lesions increased sharply after the fish began the fresh water phase of their migration. Among 69 trout taken upstream en route to the spawning beds, 63 had abnormal coronary vessels. Intimal changes consisted of hyperplastic intraluminal nodules. However, in 16 fish, which had survived a previous spawning migration, six had normal coronary arteries. Thus although the incidence of coronary lesions approached 100% in fish at the time of spawning, lesion regression appeared to occur after spawning consistent with the observations of Maneche et al. (1972).

Moore et al. (1976a and b) reported on the number, location and severity of coronary artery lesions in spawning Pacific salmon and Steelhead trout. Pacific salmon after hatching in fresh water spend three to five years in the ocean before returning to fresh water to spawn. Pacific salmon do not survive spawning. Consistent with the studies of Maneche et al. (1972) and Van Citters and Watson (1968), the incidence and severity of coronary lesions increased with maturation of the fish. Again lesions consisted of focal areas of proliferation of smooth muscle cells projecting into the lumen through a broken or split elastic lamina. These same investigators (Moore et al., 1976c) also examined the ultrastructure of coronary lesions in Pacific salmon and Steelhead trout. Lesions consisted of focal proliferation of smooth muscle cells. There was no elastic lamina below the endothelium in the lesion. Smooth muscle cells in the lesions were oriented in different directions from the cells in the medial layer of the artery. The endothelium overlying the lesion had large vacuoles and a rough irregular lower border. In fact, they observed large vacuoles in the endothelium of fish with many lesions. On the basis of this observation, they suggested that a change in the relation of the endothelium to the elastic lamina such that close contact between them is lost is somehow involved in the genesis of lesions. No lipid bodies were noted in any cell.

In a more recent study, House and Benditt (1981) also examined the ultrastructure of coronary artery lesions in Steelhead trout. The trout coronary artery is a muscular artery and the endothelium normally rests directly upon the internal elastic lamina. Most lesions were focal and contained proliferations of smooth muscle cells accompanied by extracellular collagen and elastin and capped by an intact endothelium. These smooth muscle cells were smaller than those of the adjacent media. The significance of this observation is unclear. Are these cells modified medial smooth muscle cells or are they intimal in origin? The internal elastic lamina was thinned or duplicated but not absent as reported by Moore et al. (1976c). There was no evidence of platelet or macrophage attachment to the endothelium. Although the endothelium was always intact, endothelial cells with large vacuoles were seen in association with most lesions consistent with the observation of Moore *et al.* (1976c). No lipid was detected in any lesion.

The observation of Maneche *et al.* (1972) and Van Citters and Watson (1968) that coronary lesions regress after spawning has not been confirmed by several recent studies. Saunders and Farrell (1988) examined the coronary arteries of 209 Atlantic salmon at various stages of recovery after spawning. The histological examination performed by Saunders and Farrell was much more extensive than that used in the studies of Maneche *et al.* (1972) and Van Citters and Watson (1968). There was little evidence for lesion regression; the incidence and severity of lesions examined at various times after spawning were not reduced in comparison with salmon at or near spawning condition. With respect to Steelhead trout, Farrell and Johansen (1992) found a high prevalence and severity of coronary lesions in wild repeat-spawning Steelhead trout that were caught at high sea and in wild and cultured Steelhead trout that had been held in sea pens for up to one year after maturation. Farrell and Johansen (1992) refuted the idea of natural lesion regression in Steelhead trout.

The severity and prevalence of coronary lesions in mature salmon has been well documented by Farrell *et al.* (1990). Of 221 mature Pacific salmonids sampled from five species, no fish was found to be completely free of lesions. Serial histological examinations showed that between 66% and 80% of the length of the main coronary artery contained some form of severe lesion (myointimal hyperplasia with splitting, fragmentation or loss of the elastic membrane). These severe lesions would typically occlude 10% to 30% and as much as 50% of the vessel lumen.

Finally, the (non-migratory) land-locked freshwater adult rainbow trout displays coronary artery focal intimal thickenings of a similar frequency and severity as those of the (migratory) adult steelhead trout (McKenzie *et al.*, 1978). This observation suggests that factors associated specifically with life at sea play at most, a secondary role in coronary lesion development.

In summary, focal intimal thickenings consisting of intimal proliferation of vascular smooth muscle cells with a disrupted elastic lamina have been well described in salmon and trout. The accumulation of coronary lesions is cumulative over the life of the fish, i.e. an age related progression rather than an association with specific events such as sexual maturity or senescence (Farrell, 2002). In a developmental analysis of the Steelhead trout, Kubasch and Rourke (1990) documented the presence of lesions in the main coronary artery in all stages of the life cycle of this species. Once fish became sexually mature, the incidence of lesions was 100% and the severity of lesions was at its peak. In the next section, we will review what are believed to be the factors involved in the genesis of these lesions.

Genesis of Fish Lesions

Saunders *et al.* (1992) carried out a comprehensive quantitative analysis of coronary histology for all life stages of Atlantic salmon both wild-caught and hatchery-raised. They found a clear direct positive relationship between fish length (a measure of growth) and the prevalence and severity of coronary lesions in both wild and cultured salmon sampled at various life stages, from juveniles in fresh water to smolts and post-smolts beginning their marine life and the later marine feeding stages leading to return migration, spawning and post-spawning recovery. Thus as salmon grow older and bigger they accumulate more lesions. These data are inconsistent with either sexual maturation or senescence being primary factors in the initiation of coronary lesions in salmon. The authors discuss the possible roles of diet, long-term endocrinological changes and metabolic adjustments associated with rapid growth during the marine stage in the genesis/progression of coronary lesions. However, a more attractive hypothesis is that the genesis/progression of lesions reflects the primary influence of biomechanical factors as discussed in the next section. According to the authors, periods of rapid growth are associated with episodes of hypertension, which would accentuate the biomechanical stresses responsible for lesion initiation/progression. Currently, diet and metabolic adjustments are considered secondary factors in lesion initiation and progression. The role of diet and serum lipids in lesion genesis will be considered in a later section.

Biomechanical Factors in Fish

The coronary lesions described in salmon and trout are generally found along the length of the main subepicardial coronary artery (Farrell, 2002; Moore *et al.*, 1976a and b). In fact, most sections of the coronary are generally taken cephalad of the first bifurcation of the artery (Saunders *et al.*,

1992). Such is not the case for elasmobranchs. Farrell *et al.* (1992) examined portions of the main coronary artery in five species of elasmobranchs. There was a complete lack of lesions in all of the five species. However, elasmobranchs do indeed have coronary artery thickenings and these lesions have a morphology similar to those of the salmon and trout. Garcia-Garrido *et al.* (1993) reported that in the dogfish the large subepicardial coronary arteries were relatively unaffected by lesions except at branch points. Among 90 specimens in which subepicardial arteries were sampled, only 13 specimens showed moderate lesions (one or a few small intimal nodules) along the length of the vessel. However all subepicardial coronary artery branchings showed breakage, replication or disappearance of the inner elastic layer and diffuse intimal thickening (hyperplasia). Unlike the salmon in which the coronary circulation supplies the compact myocardium only, elasmobranchs have a coronary circulation in both the compact and spongy myocardium (see "Fish Coronary Circulation" section). The intramyocardial ventricular branches in the dogfish (supplying the spongy myocardium) showed extensive lesions not associated with branching points. In some instances the lesions were so severe as to occlude most of the lumen.

What is the explanation for the difference in distribution of coronary lesions between the salmonid (salmon and trout) and the elasmobranch? In the dogfish, intimal thickenings were present at branching points of the main coronary arteries but very scarce elsewhere along these vessels. By contrast, the main coronary arteries of the salmon and trout are severely affected along their course but not particularly at branchings. The explanation lies in an important anatomical difference between elasmobranchs and salmonids. The chamber into which the ventricle empties differs functionally between these two species of fish (Farrell *et al.*, 1992). In the elasmobranch, this chamber (conus arteriosus) contains cardiac muscle which rhythmically contracts with each heart beat. In addition, the conus is valved and this valving limits the extent to which the conus is physically distended when arterial blood pressure is elevated. In the salmonid, this chamber (bulbus arteriosus) does not contain cardiac muscle or valves beyond the bulbo-ventricular orifice. In contrast to the conus, the elastic bulbus functions to depulsate blood flow by expanding with each ejection of blood from the ventricle. Elevations of arterial blood pressure cause an even larger distention of the bulbus. Based on these characteristics, mechanical deformation of the bulbus during blood ejection from the ventricle would produce mechanical stresses on the coronary artery, which lies on its surface.

The cumulative effect of these mechanical stresses would lead to injury of the coronary artery and lesion formation. The more gentle deformation of the elasmobranch conus would produce a lower stress on the arterial wall. Coronary lesions in the elasmobranch are very rare in subepicardial coronary arteries except at their branch points. However, in the elasmobranch, intramyocardial vessels do have lesions along their course (not associated with branching points) presumably because these vessels are subjected to stress and vibrations derived from both the pulsatile blood flow between atrium and ventricle and the actions of the atrioventricular valve leaflets (Garcia-Garrido et al., 1993). This biomechanical hypothesis is supported by in vitro experimental results. Gong and Farrell (1995) using rainbow trout gently rubbed the outside wall of the main coronary artery, which lies on the bulbus arteriosus and ventral aorta. One to three days post-abrasion, the abraded portion of the coronary artery was removed and an arterial explant was isolated. The bulk of the explant was vascular smooth muscle cells. Radioactive thymidine incorporation into the smooth muscle cells was used as an index of mitotic activity. Gentle coronary artery abrasion resulted in a substantial increase in vascular smooth muscle cell mitotic activity compared to coronary artery explants from sham-operated and untreated control groups of fish. If biomechanical factors are the primary determinants of coronary artery lesions, then it follows that stressful events that cause blood pressure elevations in salmon with over-expansion of the bulbus should accentuate lesion formation. Consistent with this notion is the observation that swimming to exhaustion, as a result of a critical swimming speed test, significantly increased vascular smooth muscle cell mitosis in coronary artery explants (Farrell, 2002). In contrast, continuous low speed swimming for three months had no significant effect on vascular smooth muscle cell mitosis in coronary explants.

Dietary Influences and Role of Cholesterol in Fish

The experimental observations currently available suggest that diet and cholesterol probably play a secondary role in lesion formation and progression.

Eaton et al. (1984) examined the relationship between plasma lipoprotein concentration and histological myointimal proliferative lesions in the coronary vessels of mature freshwater Chinook salmon. The fish were obtained during the migrating, schooling and spawning periods of the fourth year of life. The incidence and severity of lesions in each size vessel was greatest in the migrating salmon and progressively decreased as the salmon moved through the stages of schooling and spawning. Total serum cholesterol was high in these fish (517 to 580 mg/dl) and did not differ between the three populations. However there was a progressive increase in ApoA polypeptide serum concentration and a progressive decrease in ApoB polypeptide serum concentration in going from migrating to schooling to spawning salmon. For example, in migrating salmon mean serum concentrations were: total cholesterol 517 mg/dl, ApoA 392 mg/dl and ApoB 387 mg/dl. In spawning salmon the mean serum concentrations were: total cholesterol 580 mg/dl, ApoA 501 mg/dl and ApoB 93 mg/dl. ApoA and ApoB are the major apoproteins of HDL and LDL, respectively. Thus there was a positive correlation between lesion prevalence and severity and ApoB concentration and a negative correlation between lesion prevalence and severity and ApoA concentration. In these salmon then, excessive exposure to high concentrations of ApoB containing lipoproteins is associated with a greater prevalence and severity of focal coronary artery intimal thickenings. However, even in the spawning subgroup, with the lowest serum ApoB concentration and the highest ApoA serum concentration, the incidence of lesions in large vessels was still 38% compared to 57% in migrating salmon. In the migrating salmon the group with the highest serum ApoB concentration, no lipid deposits were described in the coronary artery lesions. Farrell *et al.* (1986) examined coronary vessel histology in cultured Atlantic salmon maintained on normal and cholesterol enriched diets during the period when they normally mature. Fish were categorized as being either immature or mature and were maintained in a freshwater or saltwater environment. Fish fed the 3% cholesterol supplement had significantly higher total plasma cholesterol levels than fish on the control diet. For example, mature male fish maintained in freshwater eating a normal diet had mean plasma concentrations of total cholesterol 450 mg/dl, HDL 390 mg/dl and LDL 60 mg/dl. When a matched group of fish was fed a 3% cholesterol supplement mean plasma concentrations were: total cholesterol 710 mg/dl, HDL 550 mg/dl and LDL 160 mg/dl. This fish has a typically high HDL concentration (about 80% or greater of total plasma cholesterol concentration) but the cholesterol-enriched diet did lead to an increase in LDL concentration of between two to five fold. In some subgroups such as immature male fish kept in freshwater, LDL levels as high as 250 mg/dl were observed. The extent of lesion formation was high in all groups and almost 95% of fish had at least one identifiable lesion. Many lesions were severe and produced more than 50% luminal narrowing. No fatty deposits were observed in the lesions. The high lesion incidence in all subgroups meant that the possible effects of diet and maturation were superimposed on a high background frequency of lesions. Nevertheless fish fed the cholesterol-enriched diet had a greater incidence and severity of lesions than fish fed the control diet. Likewise lesion incidence and severity were higher in mature compared to immature fish. These experimental results implicate diet and maturation as secondary factors in the development of coronary artery lesions in Atlantic salmon.

The study by Garcia-Garrido *et al.* (1993) on coronary artery lesions in the dogfish has already been discussed in the "Biomechanical Factors" section. Serum lipid concentrations in this elasmobranch are low with a mean total cholesterol concentration of 96 mg/dl and a mean HDL cholesterol concentration of 8.3 mg/dl. Garcia-Garrido *et al.* found no significant statistical correlation between serum lipid concentrations (total cholesterol and HDL cholesterol) and coronary artery lesion incidence and severity.

In summary, current data suggest that elevated cholesterol concentrations, probably LDL, are associated with an increase in the prevalence and severity of coronary artery lesions but are not the primary initiating factor. In addition, foam cells and lipid deposits are not observed in coronary artery focal thickenings even in fish with LDL levels as high as 250 mg/dl. O'Keefe *et al.* (2004), based upon cholesterol concentrations in hunter-gatherer populations and in wild mammals, have proposed that in humans target levels for total cholesterol and LDL to minimize atherosclerosis and coronary heart disease events should be less than 150 mg/dl and 70 mg/dl, respectively. In the study of Farrell *et al.* (1986) most of the fish populations fed the cholesterol supplement had LDL levels between 150 and 280 mg/dl, significantly above the target levels proposed by O'Keefe *et al.* (2004). Why do advanced complex mammalian type atherosclerotic lesions not develop in the coronary artery focal thickenings of these fish? Is it because the high HDL level found in these fish counteracts the atherogenic influence of an elevated LDL concentration?

The possible influence of omega 3 polyunsaturated fatty acids (PUFA) on coronary artery lesion prevalence requires some discussion. Juvenile Atlantic salmon in their fresh water streams feed largely on aquatic insects in which the predominant PUFA is of the omega 6 type. Marine fish, which comprise a large portion of the diet when salmon are maturing in the sea, are rich in omega 3 PUFAs. In a six-month feeding trial in salmon using diets rich in omega 3 or omega 6 PUFAs, there was no evidence that diets enriched with omega 3 PUFAs led to any changes in the progression of coronary lesions (Saunders *et al.*, 1992). It appears then that the shift from a high omega 6 to a high omega 3 diet has no major impact on lesion development and that, moreover the high dietary omega 3 intake during the sea phase of the life cycle does not prevent the rapid development of coronary lesions in Atlantic salmon (Saunders *et al.*, 1992).

Biology of the Vascular Smooth Muscle in Fish

Even though biomechanical factors appear to be primary in initiating coronary artery lesions in fish, there still needs to be a metabolic signaling mechanism linking the biomechanical factors to the response, i.e. proliferation of vascular smooth muscle cells. In this regard, PUFAs seem to be important regulators of smooth muscle cell proliferation in salmon. Gong *et al.* (1997) studied the effects of PUFAs on mitotic activity of vascular smooth muscle explants from the coronary artery of rainbow trout. The effects are rather complex and only a few are described here. For example, a low dose of arachidonic acid in the culture medium caused a five fold stimulation of smooth muscle mitosis. The omega 3 PUFA eicosapentaenoic acid (EPA) by itself had little effect on smooth muscle cell mitosis but inhibited the mitogenic effect of arachidonic acid. According to Gong *et al.* (1997) these effects are similar to those observed for mammalian vascular smooth muscle cells. Arachidonic acid induces vascular smooth muscle cell proliferation in aortic vascular smooth muscle cell cultures from guinea pigs and mice. However aside from these studies on vascular smooth muscle cell proliferation, there are no data as to whether salmon vascular smooth muscle cells are capable of major changes in phenotypic expression, and if so what factors might regulate these changes.

Concluding Remarks: What Can We Learn From the Fish as a Natural Model?

Can an understanding of the formation and biology of coronary artery lesions in the salmon, trout and dogfish help us in our understanding of the pathogenesis of atherosclerosis in humans? In the human, focal intimal thickenings or cushions develop *in utero* in the coronary vessels and in early infancy in the internal carotid artery. Less extensive focal intimal cushions have also been described in the basilar, anterior cerebral, renal, hepatic and bronchial arteries. The primary initiating determinant is most likely biomechanical either hemodynamic/blood flow patterns or perhaps shape alterations in the vessel. Biomechanical factors would explain the distribution of these cushions at branch points (coronary artery) or bends (internal carotid artery). The apparent susceptibility of the coronary arteries and the parasellar internal carotid artery to the development of focal intimal thickenings or cushions probably relects the extreme hemodynamic/biomechanical stresses acting on these vessels. These intimal cushions provide the soil/site for the future development of atherosclerotic lesions although the latent period is many years. In the fish, focal intimal thickenings in the coronary arteries histologically resemble these intimal cushions in humans. In the fish, the most likely initiating factor is also biomechanical as a result of the stresses the coronary arteries experience as described in the "Biomechanical Factors" section. The widespread prevalence of these lesions in the fish would strongly suggest that they are adaptive and probably protect the integrity of the coronary vessels from the biomechanical stresses associated with cardiac contractions and episodes of hypertension. However these lesions can be quite severe and in some cases may almost occlude the vessel lumen. The functional consequences of these severe lesions are not clear and in particular it is not known if they actually limit fish survival and if so under what circumstances. This particular issue is discussed by Farrell (2002). By analogy with the situation in the fish are the intimal cushions in the human coronary and internal carotid arteries also adaptive? Do these focal intimal thickenings develop as in the fish, to protect the integrity of these vessels?

Stehbens (1996) has discussed this question. He argues that in the human arterial tree, hemodynamic stresses result in bioengineering fatigue of the vessel wall. This process begins *in utero* and occurs at branch points. Progressive disruption of the internal elastic and medial elastic lamina weakens the vessel wall. Stehbens views the proliferation of smooth muscle cells (i.e. formation of a focal intimal thickening) as a compensatory or reparative process to maintain vessel integrity and architecture. Hence, the focal arterial intimal thickenings in the fish and human appear to be comparable. The formation of these intimal thickenings represents a shared response designed to maintain arterial integrity in the face of chronic repetitive hemodynamic/biomechanical stresses. There is however, a significant biological difference between fish and human intimal thickenings.

In the human, intimal thickenings provide a microenvironment or soil for the development of atherosclerotic lesions which presumably occur as a result of the sequence of events described in the "Atherosclerosis in Mammals" section. In the fish intimal thickenings do not become the site for the characteristic atherosclerotic changes observed in humans. In the case of the fish, lesion progression refers to the extent of vascular smooth muscle cell proliferation. A greater amount of proliferation results in more extensive intimal thickening with projection of the intimal nodule into the vessel lumen with partial lumen occlusion. There is no evidence of macrophage or platelet attachment to the endothelium and lipid deposits are not observed in the intimal thickenings. Why do the intimal thickenings in fish not develop into the advanced complicated atherosclerotic lesions found in mammals or more particularly humans? Unfortunately currently available data do not give a definitive answer to this question but do allow for certain speculations. Simplistically there are several possibilities.

The microenvironment of the fish intimal thickenings could differ from that of the human such that atherosclerotic lesions do not develop or the fish could lack risk factors, which in the human are strongly associated with the atherosclerotic process. Perhaps the most dominant risk factor in humans is hyperlipidemia and in particular elevated serum concentrations of LDL and low serum concentrations of HDL. LDL is considered to play a pivotal role in the genesis of atherosclerotic lesions as discussed in the "Atherosclerosis in Mammals" section. The relationship of serum lipid concentrations to the prevalence and severity of fish intimal thickenings is discussed in the "Dietary Influences and Role of Cholesterol" section. Fish fed an atherogenic high cholesterol diet do have a greater incidence and severity of intimal thickenings but no mammalian type atherosclerotic changes develop. However, in salmonids, HDL levels are very high (about 80% of the total serum cholesterol concentration) and this high HDL level might mitigate the effects of the elevated serum LDL concentration induced by the atherogenic diet. Hypertension is another risk factor in humans. In salmon, periods of rapid growth are believed to be associated with hypertension, and as discussed in the "Genesis of Fish Lesions" section, fish length (a measure of growth) directly correlates with the prevalence and severity of lesions. Hence, as salmon grow older and bigger they accumulate more lesions but none display mammalian type atherosclerotic changes. Hence, the very limited data suggest that risk factors such as hyperlipidemia and hypertension act as secondary factors to increase only the severity and incidence of fish arterial intimal thickenings. A reasonable speculation is that the microenvironment within fish intimal thickenings is resistant to the series of cellular and molecular responses that result in the formation of the advanced complicated atherosclerotic lesions in humans. If this speculation is correct, it means that there is a biological solution to the problem of atherosclerosis. The underlying basis for such a "resistance" could reflect special characteristics of fish endothelial and vascular smooth muscle cells, the two cell types comprising the intimal thickening. If this speculation is considered a working hypothesis, then a number of research questions follow. How do the biological properties of fish endothelial and vascular smooth muscle cells compare to those of the mammal/human? Why are monocytes and lymphocytes not recruited to the arterial luminal surface in the fish and why are foam cells not found in fish intimal thickenings? The respective roles of LDL and HDL in the progression of fish coronary thickenings are unknown. Even in fish with high LDL levels, mammalian type atherosclerotic lesions do not develop perhaps because of the counteracting influence of high HDL concentrations or perhaps because the microenvironment of the fish focal thickening is resistant to the uptake of LDL into the subendothelial space. Could genetic techniques such as creation of a fish comparable to the homozygous mutant ApoE deficient mouse be used to distinguish between these two possibilities and to dissect out the individual roles of HDL and LDL in the progression of fish coronary lesions? If intimal thickenings in the fish do not provide a suitable soil for the development of atherosclerotic lesions, then understanding the differential properties between fish and human arterial intimal thickenings could lead to strategies aimed at either preventing atherosclerosis or at least intervening at a very early stage.

References

Burggren, W.W., 2004. What is the purpose of the embryonic heart beat? Or how facts can ultimately prevail over physiological dogma. *Physiol. Biochem. Zool.* 77, 333–345.

Clark, J.J., Clark, R.J., McMinn, J.T., Rodnick, K.J., 2004. Microvascular and biochemical compensation during ventricular hypertrophy in male rainbow trout. *Comp. Biochem. Physiol. Part B* 139, 695–703.

Davie, P.S., Farrell, A.P., 1991. The coronary and luminal circulation of the myocardium of fishes. *Can. J. Zool.* 69, 1993–2001.

Doherty, T.M., Shah, P.K., Rajavashisth, T.B., 2003. Cellular origins of atherosclerosis: towards ontogenetic endgame? *FASEB J.* 17, 592–597.

Eaton, R.P., McConnell, T., Hnath, J.G., Black, W., Swartz, R.E., 1984. Coronary myointimal hyperplasia in freshwater Lake Michigan salmon (genus *Oncorhynchus*). Evidence for lipoprotein-related atherosclerosis. *Am. J. Pathol.* 116, 311–318.

Farrell, A.P., Saunders, R.L., Freeman, H.C., Mommsen, T.P., 1986. Arteriosclerosis in Atlantic salmon. Effects of dietary cholesterol and maturation. *Arteriosclerosis* 6, 453–461.

Farrell, A.P., Steffensen, J.F., 1987. Coronary ligation reduces maximum sustained swimming speed in Chinook salmon, *Oncorhynchus tshawytscha. Comp. Biochem. Physiol.* 87A, 35–37.

Farrell, A.P., Johansen, J.A., Saunders, R.L., 1990. Coronary lesions in Pacific salmonids. *J. Fish. Dis.* 13, 97–100.

Farrell, A.P., Johansen, J.A., 1992. Reevaluation of regression of coronary arteriosclerotic lesions in repeat-spawning steelhead trout. *Arterioscler. Thromb.* 12, 1171–1175.

Farrell, A.P., Davie, P.S., Sparksman, R., 1992. The absence of coronary arterial lesions in five species of elasmobranchs, *Raja nasuta, Squalus acanthias* L., *Isurus oxyrinchus Rafinesque, Prionace glauca* (L.) and *Lamna nasus* (Bonnaterre). *J. Fish. Dis.* 15, 537–540.

Farrell, A.P., 2002. Coronary arteriosclerosis in salmon: growing old or growing fast? *Comp. Biochem. Physiol. Part A* 132, 723–735.

Franklin, C.E., Axelsson, M., 1994. Coronary hemodynamics in elasmobranches and teleosts. *Cardioscience* 5, 155–161.

French, J.E., 1966. Atherosclerosis in relation to the structure and function of the arterial intima, with special reference to the endothelium. *Int. Rev. Exp. Pathol.* 5, 253–353.

Garcia-Garrido, L., Munoz-Chapuli, R., de Andres, V., 1993. Coronary arteriosclerosis in dogfish (*Scyliorhinus canicula*). An assessment of some potential risk factors. *Arterioscler. Thromb.* 123, 876–885.

Gimbrone, Jr., M.A., 1999. Vascular endothelium, hemodynamic forces, and atherogenesis. *Am. J. Pathol.* 155, 1–5.

Gong, B.Q., Farrell, A.P., 1995. A method of culturing coronary artery explants for measuring vascular smooth muscle proliferation in rainbow trout. *Can. J. Zool.* 73, 623–631.

Gong, B., Townley, R., Farrell, A.P., 1997. Effects of polyunsaturated fatty acids and some of their metabolites on mitotic activity of vascular smooth muscle explants from the coronary artery of rainbow trout (*Oncorhynchus mykiss*). *Can. J. Zool.* 75, 80–86.

House, E.W., Benditt, E.P., 1981. The ultrastructure of spontaneous coronary arterial lesions in steelhead trout (*Salmo gairdneri*). *Am. J. Pathol.* 104, 250–257.

Ikari, Y., McMzanus, B.M., Kenyon, J., Schwartz, S.M., 1999, Neonatal intima formation in the human coronary artery. *Arterioscler. Thromb. Vasc. Biol.* 19, 2036–2040.

Kubasch, A., Rourke, A.W., 1990. Arteriosclerosis in steelhead trout, *Oncorhynchus mykiss* (Walbaum): a developmental analysis. *J. Fish. Biol.* 37, 65–69.

Li, D.Y., Brooke, B., Davis, E.C., Mecham, R.P., Sorensen, L.K., Boak, B.B., Eichwald, E., Keating, M.T., 1998. Elastin is an essential determinant of arterial morphogenesis. *Nature* 393, 276–280.

Libby, P., 2002. Inflammation in atherosclerosis. *Nature* 420, 868–874.

Lusis, A.J., 2000. Atherosclerosis. *Nature* 407, 233–241.

Maneche, H.C., Woodhouse, S.P., Elson, P.F., Klassen, G.A., 1972. Coronary artery lesions in Atlantic salmon (*Salmo salar*). *Exp. Mol. Pathol.* 17, 274–280.

McKenzie, J.E., House, E.W., McWilliam, J.G., Johnson, D.W., 1978. Coronary degeneration in sexually mature rainbow and steelhead trout, *Salmo gairdneri*. *Atherosclerosis* 29, 431–437.

Moore, J.F., Mayr, W., Hougie, C., 1976a. Number, location and severity of coronary arterial changes in spawning Pacific salmon (*Oncorhynchus*). *J. Comp. Pathol.* 86, 37–43.

Moore, J.F., Mayr, W., Hougie, C., 1976b. Number, location and severity of coronary arterial changes in steelhead trout (*Salmo gairdnerii*). *Atherosclerosis* 24, 381–386.

Moore, J.F., Mayr, W., Hougie, C., 1976c. Ultrastructure of coronary arterial changes in spawning Pacific salmon (genus *Oncorhynchus*) and steelhead trout (*Salmo gairdnerii*). *J. Comp. Path.* 86, 259–267.

Mulvihill, E.R., Jaeger, J., Sengupta, R., Ruzzo, W.L., Reimer, C., Lukito, S., Schwartz, S.M., 2004. Atherosclerotic plaque smooth muscle cells have a distinct phenotype. *Arterioscler. Thromb. Vasc. Biol.* 24, 1283–1289.

Nakashima, Y., Plump, A.S., Raines, E.W., Breslow, J.L., Ross, R., 1994. Apo E-deficient mice develop lesions of all phases of atherosclerosis throughout the arterial tree. *Arterioscler. Thromb.* 14, 133–140.

Ohashi, R., Mu, H., Yao, Q., Chen, C., 2004. Atherosclerosis: immunopathogenesis and immunotherapy. *Med. Sci. Monit.* 10, RA255–RA260.

O'Keefe, Jr., J.H., Cordain, L., Harris, W.H., Moe, R.M., Vogel, R., 2004. Optimal low-density lipoprotein is 50 to 70 mg/dl. Lower is better and physiologically normal. *J. Am. Coll. Cardiol.* 43, 2142–2146.

Osterud, B., Bjorklid, E. 2003. Role of monocytes in atherogenesis. *Physiol. Rev.* 83, 1069–1112.

Owens, K., Kumar, M.S., Wamhoff, B.R., 2004. Molecular regulation of vascular smooth muscle cell differentiation in development and disease. *Physiol. Rev.* 84, 767–801.

Robertson, O.H., Krupp, M.A., Thomas, S.F., Favour, C.B., Hane, S., Wexler, B.C., 1961. Hyperadrenocorticism in spawning migratory and non-migratory rainbow trout (*Salmo gairdnerii*); comparison with Pacific salmon (Genus *Oncorhynchus*). *Gen. Comp. Endocrinol.* 1, 473–484.

Ross, R., Glomset, J.A., 1973. Atherosclerosis and the arterial smooth muscle cell. *Science* 180, 1332–1339.

Ross, R., 1993. The pathogenesis of atherosclerosis: a perspective for the 1990s. *Nature* 362, 801–809.

Ross, R., 1999. Atherosclerosis — an inflammatory disease. *N. Engl. J. Med.* 340, 115–126.

Saunders, R.L., Farrell, A.P., 1988. Coronary arteriosclerosis in Atlantic salmon. No regression of lesions after spawning. *Arteriosclerosis* 8, 378–384.

Saunders, R.L., Farrell, A.P., Knox, D.E., 1992. Progression of coronary arterial lesions in Atlantic salmon (*Salmo salar*) as a function of growth rate. *Can. J. Fish. Aquat. Sci.* 49, 878–884.

Schwartz, S.M., deBlois, D., O'Brien, E.R.M., 1995. The intima. Soil for atherosclerosis and restenosis. *Circ. Res.* 77, 445–465.

Schwartz, S.M., 1999. The intima. A new soil. *Circ. Res.* 85, 877–879.

Scott, R.F., Thomas, W.A., Lee, W.M., Reiner, J.M., Florentin, R.A., 1979. Distribution of intimal smooth muscle cell masses and their relationship to early atherosclerosis in the abdominal aortas of young swine. *Atherosclerosis* 34, 291–301.

Stary, H.C., 1989. Evolution and progression of atherosclerotic lesions in coronary arteries of children and young adults. *Arteriosclerosis* 9, Supp. 1, 1–19 to 1–32.

Stary, H.C., Blankenhorn, D.H., Chandler, A.B., Glagov, S., Insull Jr., W., Richardson, M., Rosenfeld, M.E., Schaffer, S.A., Schwartz, C.J., Wagner, W.D., Wissler, R.W., 1992. A definition of the intima of human arteries and of its atherosclerosis-prone regions. *Circulation* 85, 391–405.

Stehbens, W.E., 1996. Structural and architectural changes during arterial development and the role of hemodynamics. *Acta Anat.* 157, 261–274.

Van Citters, R.L., Watson, N.W., 1968. Coronary disease in spawning steelhead trout (*Salmo gairdnerii*). *Science* 159, 105–107.

Velican, C., Velican, D., 1976. Intimal thickening in developing coronary arteries and its relevance to atherosclerotic involvement. *Atherosclerosis* 23, 345–355.

Velican, C., Velican, D., 1979. Some particular aspects of the microarchitecture of human coronary arteries. *Atherosclerosis* 33, 191–200.

Wada, A.M., Willet, S.G., Bader, D., 2003. Coronary vessel development. A unique form of vasculogenesis. *Arterioscler. Thromb. Vasc. Biol.* 23, 2138–2145.

Weninger, W.J., Muller, G.B., Reiter, C., Meng, S., Rabl, S.V., 1999. Intimal hyperplasia of the infant parasellar carotid artery. A potential developmental factor in atherosclerosis and SIDS. *Circ. Res.* 85, 970–975.

Zhang, S.H., Reddick, R.L., Piedrahita, J.A., Maeda, N., 1992. Spontaneous hypercholesterolemia and arterial lesions in mice lacking Apolipoprotein E. *Science* 258, 468–471.

Disuse Osteoporosis and Disuse Muscle Atrophy

In this chapter, two clinical disorders, disuse osteoporosis and disuse muscle atrophy, are discussed. The natural animal model for both of these disorders is the American black bear (*Ursus americanus*). During five to seven months of winter dormancy, this animal suffers far less disuse osteoporosis and muscle atrophy than a human subjected to a comparable period of immobilization. The bear is in many ways a metabolic marvel. As reviewed in Chapter 2, the dormant bear does not eat, drink, defecate or urinate and at the same time is essentially immobile. Its physiological state can be characterized as equivalent to a combination of fasting and chronic renal failure. The adaptations described in Chapter 2 and those described in this chapter should in reality not be considered as separate or isolated from each other. These sets of adaptations are really parts of an integrated whole. However, given our incomplete understanding of the bear's metabolic characteristics, the adaptations discussed in Chapter 2 and in this chapter, have been treated as if they were separate.

Disuse Osteoporosis

Before discussing what is known about the bear's adaptations to a prolonged period of inactivity, some relevant aspects of mammalian bone
biology and what happens in other mammals when immobilized for an
extended time are reviewed.

Bone biology

Bone is a vital dynamic connective tissue which has two major functions:
load bearing (to support posture, permit movement including locomotion
and provide protection for soft tissues) and involvement in the metabolic
processes associated with calcium homeostasis (Lee and Einhorn, 2001). To
fulfill these functions, bone is constantly being broken down and rebuilt in a
process referred to as remodeling. Bone is a composite material containing
both an organic and inorganic phase. About 95% of the inorganic phase
(mineral) phase is composed of a specific crystalline hydroxyapatite with
the other 5% consisting of imbedded impurities. Ninety-eight percent of
the organic phase is composed of collagen (predominantly type 1) and a
variety of non-collagenous proteins; cells make up the other 2% of this
phase.

The skeleton consists of two parts: an axial skeleton which includes the
vertebrae, pelvis and other flat bones (e.g. skull and sternum) and an appendicular skeleton which includes all of the long bones (Lee and Einhorn,
2001). Anatomically bone is organized into trabecular (cancellous) and cortical (compact) compartments. Cortical bone has four times the mass of
trabecular bone but since trabecular bone has a greater surface area than
cortical bone, it has a higher metabolic turnover rate. In general, cancellous
bone is the main tissue of the axial skeleton whereas cortical bone is the
main tissue of the appendicular skeleton. Cancellous or trabecular bone
is generally subject to a complex set of stresses and strains although it is
best designed for resisting compressive loads. Microscopically it consists
of plates which form a three-dimensional branching lattice which is oriented along lines of stress. Cortical bone is usually subject to bending and
torsional forces as well as compressive loads. Cortical bone is solid and
arranged microscopically as cylinders. Following skeletal maturity, bone
continues to remodel throughout life and adapt its material properties to
the prevailing mechanical demands. Remodeling is a continuous process
orchestrated by osteoblasts and osteoclasts on bone surfaces.

The osteoblast or bone forming cell has as its primary functional activity the production of an extracellular matrix with properties conducive to mineralization (Lian and Stein, 2001). The origin of the osteoblast lineage cells has not been determined. Alkaline phosphatase activity, a hallmark of the osteoblast phenotype, is a widely accepted marker of new bone formation and early osteoblast activity. Cells of the osteoblast lineage express bone matrix proteins: collagen (chiefly type 1), glycosaminoglycan-containing proteins and glycoproteins to name a few of the types. As the active matrix forming osteoblast becomes encased in the mineralized matrix, the cell differentiates further into an osteocyte, the terminally differentiated cell of the osteoblast lineage. Osteocytes can both synthesize new bone matrix at the surface of the spaces they occupy as well as resorbing calcified bone from the same surface. Osteocytes appear to sense mechanical forces and there is evidence that mechanical tension can trigger bone remodeling and may favor bone formation (Lian and Stein, 2001). The inorganic phase of bone is composed mainly of a calcium phosphate mineral analogous to crystalline calcium hydroxyapatite. This apatite is present as a plate-like crystal. Small amounts of impurities such as carbonate can replace phosphate groups. Cells of the osteoblast lineage regulate bone matrix mineralization as reviewed by Gokhale *et al.* (2001)

The osteoclast is the prinicipal, if not exclusive resorptive cell of bone (Ross and Teitelbaum, 2001). The osteoclast is hematopoietic in origin belonging to the monocyte/macrophage family. The initial step in bone resorption is attachment of the osteoclast to the matrix followed by the creation of an isolated extracellular resorptive microenvironment. Dissolution of the inorganic phase precedes that of protein. Demineralization involves acidification of the extracellular microenvironment, a process mediated by vacuolar H^{+-}ATPase in the ruffled membrane of the osteoclast. Acidification of the isolated resorptive environment leads to mobilization of mineral as well as subsequent solubilization of the organic phase of bone. The products are endocytosed by the osteoclast and transported to and released at its antiresorptive surface. Some of the factors regulating bone resorption include Vitamin D and parathyroid hormone (PTH) whose actions will be discussed in a subsequent section.

Bone remodeling refers to the renewal process whereby small pockets of old bone disposed throughout the skeleton are replaced by new bone throughout adult life (Martin and Rodan, 2001). Remodeling is thus a continuous process involving a constant removal and replacement of whole volumes of bone tissue. Remodeling is essential for the maintenance of normal bone structure and for calcium homeostasis. It occurs only on bone surfaces and involves a complex interaction between osteoblast lineage cells and osteoclasts (Martin and Rodan, 2001). Stimuli initiating remodeling include circulating hormones, locally generated regulatory factors and mechanical forces. The effect of mechanical forces on bone formation and resorption is well documented (Martin and Rodan, 2001). For example, a decrease in mechanical load produced by immobilization causes a reduction in bone mass which is due both to increased bone resorption which occurs initially and to decreased bone formation which is sustained for a longer duration. Eventually the system reaches a new steady state where the available bone mass is probably adequate for the prevailing mechanical load.

The normal remodeling sequence is believed to be as follows: quiescence, activation, resorption, reversal, formation and return to quiescence (Lee and Einhorn, 2001). The conversion of an area of bone surface from quiescence to activity is referred to as activation. Activation involves recruitment and attachment of osteoclasts to the bone surface. Resorption then occurs. The reversal phase is the time interval between the completion of resorption and the initiation of bone formation at a particular site. Osteoblasts appear at the base of the resorption cavity. Unlike resorption, bone formation is a two-step process. First osteoid is synthesized and laid down at specific sites. Following this, the osteoblasts must mineralize the newly formed protein matrix. Remodeling is an integrated process involving coupling between resorption and formation. The mediators which regulate bone remodeling and bone mass homeostasis have been reviewed by Martin and Rodin (2001). As already noted, they include circulating hormones, locally generated regulatory factors and mechanical forces. Somewhat arbitrarily, only four of these; parathyroid hormone (PTH), calcitriol, leptin and mechanical forces, are briefly discussed.

Parathyroid hormone (PTH)

The parathyroid gland secretes PTH in response to decreases in blood ionized calcium levels in order to maintain a normocalcemic state (Nissenson, 2001). The exquisitely sensitive control mechanism by which minute changes in serum calcium induce large changes in PTH secretion is based upon the presence of a transmembrane calcium receptor on the chief cells of the parathyroid gland (Rodriguez et al., 2005). PTH secretion is inhibited via the calcium receptor when it is activated by an elevated concentration of ionized calcium. Conversely, a low serum level of ionized calcium reduces activation of the receptor and PTH secretion increases. Although calcium (via the calcium receptor) is the primary mechanism regulating PTH secretion, blood phosphorus levels can also regulate the secretion of PTH. In animal and *in vitro* studies, high phosphorus levels stimulate PTH secretion (Rodriguez *et al.*, 2005). The mechanisms by which phosphorus regulates the secretion of PTH include a specific sodium-phosphate co-transporter in parathyroid tissue which may serve to detect changes in extracellular phosphate concentration (Rodriguez *et al.*, 2005). There is also a complex interaction between PTH and calcitriol. Calcitriol acts as a negative regulator of PTH synthesis and secretion as well as parathyroid cell proliferation (Rodriguez *et al.*, 2005). Calcitriol also regulates the expression of calcium receptor mRNA. PTH has a number of physiological actions: mobilization of calcium from bone by promoting osteoclastic bone resorption, increased bone formation through a direct action on osteoblast lineage cells, reduced renal calcium excretion, increased renal phosphorus excretion and increased production of the active form of vitamin D (calcitriol) by the kidney (Rodriguez *et al.*, 2005; Nissenson, 2001). The anabolic (bone formation) effects of PTH are discussed in more detail in the "Bone Homeostasis in the Bear: Research Questions" section.

Vitamin D

The active form of vitamin D (calcitriol; 1,25 $(OH_2)D$) is produced by sequential hydroxylations of vitamin D in the liver (25-hydroxylation) and the kidney proximal tubule (1 alpha-hydroxylation). Renal 1 alpha-hydroxylase is a tightly regulated enzyme (in contradistinction to hepatic 25-hydroxylase) and the critical determinant of calcitriol synthesis (Feldman *et al.*, 2001). The principal regulator of 1 alpha-hydroxylase is PTH. PTH significantly increases 1 alpha-hydroxylase activity in mammalian renal slices, isolated renal tubules and cultured renal cells. Phosphate (phosphorus) is the second most important physiological regulator with high phosphate (phosphorus) levels suppressing and low levels stimulating enzyme activity. The target organs of calcitriol action are intestine, bone and kidney. In the small intestine, calcitriol increases calcium and phosphate (phosphorus) absorption. The actions of calcitriol on bone are complex since this agent has effects on both osteoblasts and osteoclasts (Feldman *et al.*, 2001). Hence, calcitriol plays a role in bone formation (including mineralization) and bone resorption. In the kidney, calcitriol regulates its own production by regulating 1 alpha-hydroxylase activity. Low concentrations of calcitriol promote hydroxylase activity.

Leptin

Originally, leptin was described as a product of adipocytes that acted centrally to regulate appetite. However, leptin receptors are widely distributed and this hormone has diverse functions (Holloway *et al.*, 2002; Gordeladze *et al.*, 2002; Patel and Karsenty, 2001). Leptin appears to have multiple effects on bone homeostasis. Some experimental observations show that leptin decreases bone formation but that this effect is indirect and mediated by binding of leptin to hypothalamic receptors (Patel and Karsenty, 2001). Holloway *et al.* (2002) have shown that leptin directly inhibits the *in vitro* differentiation of cells present in the peripheral blood mononuclear cell fraction of human blood to osteoclasts and also inhibits osteoclast differentiation from mouse spleen cells. Hence, leptin is a potential local inhibitor of bone resorption *in vivo*. In addition, leptin as well as its receptor are expressed by normal human osteoblasts (Gordeladze *et al.*, 2002). *In vitro* studies with human iliac crest osteoblasts have demonstrated that leptin promotes cell proliferation, collagen synthesis, cell differentiation and (*in vitro*) mineralization (Gordeladze *et al.*, 2002). Hence, leptin can regulate bone homeostasis in multiple ways by both locally and centrally mediated mechanisms.

Mechanical stress

Stress is the internal resistance generated within bone to counteract an applied force. Strain refers to the changes in shape that bone experiences when it is subjected to an applied force. Mechanical stress is one of the determinants of bone morphology, bone mineral density and bone strength (Takata and Yasui, 2001). Osteocytes sense mechanical loading. The gap junction of the long processes of osteocytes plays an important role in transmitting mechanical load through intra- and extracellular signal transmitters (Takata and Yashui, 2001). The majority of evidence to date suggests that mechanical stress/strain can trigger bone remodeling and probably favors bone formation. Mechanical stress/strain induces factors for the proliferation, differentiation and anabolic activities of osteoblasts as well as promoting osteoclastic bone resorption as part of the remodeling process

(Lian and Stein, 2001). Bone formation/resorption induced by the homeostatic inputs of mechanical forces will be modulated by the endocrine milieu present at the time.

Disuse osteoporosis

Reduction in mechanical loading results in bone loss often referred to as disuse osteoporosis. Perhaps a better term would be immobilization osteoporosis. Immobilization can be either involuntary, as for example due to some type of paralytic condition, or may be voluntary as when volunteers are placed under continuous bed rest. Bone homeostasis has also been described in hibernating mammals other than the American black bear.

Kiratli (2001) has summarized studies reporting the bone response to paralytic injuries. These studies have involved a number of techniques: markers of metabolic bone activity (resorption and formation), histological analysis of bone biopsies and radiographic assessment of bone mass and bone mineral density. The results of these different techniques are not always mutually consistent. Metabolic markers generally show increased bone resorption as well as an increase in bone formation. On the other hand, histological evidence shows changes consistent with both increased bone loss and decreased formation. Radiographic studies demonstrate decreased bone mass although this loss is usually site specific. Some of the inconsistencies between techniques may be due to the different time lines for bone loss and bone formation following the injury and the different techniques may not be applied at the same time point post-injury. Uebelhart *et al.* (1995) reviewed bone metabolic studies in spinal cord injured individuals. There is initially a rapid and intense bone loss occurring predominantly in the paralyzed limbs associated with hypercalcemia, hypercalcuria and increased urinary excretion of hydroxyproline. Histomorphometric analysis of bone biopsies shows an increase in bone resorption surfaces as well as a reduced bone formation rate. However these changes appear to correlate with the time following the spinal cord injury. Urinary hydroxyproline excretion returns to baseline pre-injury levels about six months after the onset of the paralysis. Markers of bone formation show an initial depression of osteoblastic response but then an increase in bone formation peaking at month 7 after onset. In terms of bone mass homeostasis, major bone loss occurs during the first six months after the spinal cord injury and then stabilizes between six to 16 months at about two-thirds of original bone mass.

Voluntary immobilization studies have also been done. Leblanc *et al.* (1990) studied six healthy men confined to horizontal bed rest for 17 weeks. Bone mineral density of the calcaneus, proximal tibia, hip (femoral neck and trochanter), lumbar spine, radius, ulna and total body was measured multiple times during the period of bed rest. The total body, lumbar spine, femoral neck, trochanter, tibia and calcaneus demonstrated significant loss of bone mineral density during bed rest. Reambulation was followed for 30 weeks post-bed rest and although some areas showed an increase in bone mineral density, only the calcaneus demonstrated a statistically significant recovery. Zerwekh *et al.* (1998) studied the effects of 12 weeks of bed rest on bone homeostasis in 11 volunteers (nine men and two women). Bed rest resulted in a small but significant increase in serum calcium levels, a fall in blood levels of PTH and calcitriol and a significant increase in urinary calcium and phosphate excretion. The primary event is clearly increased bone resorption with release of calcium and elevation of the blood calcium concentration. The suppressed blood levels of PTH and calcitriol indicate that these hormones are not the prime mediators of the bone homeostatic changes during voluntary bed rest. Iliac crest bone biopsies were done immediately prior to and at the end of the 12 weeks of bed rest. Bone histology showed a marked and significant increase in the parameters of bone resorption as a result of bed rest. Both cancellous and cortical bone-eroded surfaces increased by over 100% of the pre-bed rest value. In addition, osteoblastic surface significantly declined by 39% following the 12 weeks of bed rest. During bed rest, markers of bone formation demonstrated no significant change whereas the bone resorption marker was significantly increased. Bone mass (measured as bone mineral density) decreased at the lumbar spine, femoral neck and trochanter during the period of bed rest.

In hibernating (small) mammals, bone homeostasis is characterized by significant bone loss during the hibernating period. As reviewed in Chapter 2, these small mammal hibernators exhibit periodic arousals. Hence, the hibernation season actually consists of a series of hibernation bouts. There are also data to indicate that the overall rate of protein synthesis is significantly depressed during these bouts of hibernation (Chapter 2). Mayer and Bernick (1958) studied the dentin, periodontium and alveolar bone of active versus hibernating arctic ground squirrels. After three weeks of hibernation, dentin was decalcified, alveolar bone resorbed and large periodontal pockets formed. Haller and Zimny (1977) studied mandibular alveolar bone in non-hibernating and hibernating 13-lined ground squirrels. Bone histology was studied after five to eight days of hibernation.

Bone loss occurred during the hibernating period and observations by light and transmission electron microscopy showed that the bone resorption was probably mediated by osteocytes; so called osteocytic osteolysis. Whalen *et al.* (1972) examined the cortical thickness and histology of the femur from bats (*Myotis lucifugus*) of both sexes before, during and after hibernation. These bats have a hibernation period extending from November through to early April. Measurements were made on bats captured in September, November, December, January, February and late April. A progressive thinning of the femoral cortex occurred during hibernation which reversed during the arousal stage after the hibernation season. Histologically, enhanced bone resorption was present during the hibernation season. The changes consisted of an increase in size of osteocytes and the lacunae they occupied as well as a change in the staining properties of the surrounding matrix. Osteoclasts were not observed to be involved in the progressive loss of bone tissue during hibernation. Bone morphology quickly returned to normal after arousal. Whalen *et al.* (1972) also reported that parathyroid gland function increased during the early phases of hibernation and then declined near the conclusion of hibernation. This increased parathyroid gland activity is in contrast to the reduced parathyroid activity observed during voluntary bed rest in humans. Doty and Nunez (1985) further characterized bone histological changes during hibernation and arousal in the same bat species (*Myotis lucifugus*). Adult male and female bats were caught in the wild throughout the calendar year. The femur and the tibia were the bones studied. Active bats showed a normal appearing population of hematopoietic and bone cells within the cavities of the long bones. Osteoclasts and osteoblasts were visible on the bone surface. During the hibernation season the hematopoietic cell population was absent and well differentiated osteoblasts and osteoclasts were not visible. There was some evidence that osteocytes were responsible for the bone loss during hibernation. Within 72 hours of arousal, numerous osteoclasts were present along the bone surface. Subsequently osteoblasts were also observed along the bone surfaces but only after the osteoclasts had made their appearance. The last event to occur was the hematopoietic cell repopulation of the marrow space. Thus in this mammal a repeatable sequence of events occurs during arousal consisting of the appearance of first osteoclasts, then osteoblasts and finally repopulation of hematopoietic tissues. Kwiecinski *et al.* (1987) have also examined (though in somewhat more detail) the seasonal skeletal changes in both male and female bats (*Myotis lucifugus*). Bats were collected biweekly between May to August and monthly between

September to April. Bone mineral density was lower in winter (hibernation) months than in the summer (active) months. Plasma total calcium concentrations were significantly higher during winter hibernation and peaked during January in females and February in males. For example, in the males mean calcium concentration was 9.38 mg/dl in September and peaked at 12.38 mg/dl in February. For females, the respective values were 9.74 mg/dl and 11.50 mg/dl in September and January. Plasma phosphorus concentrations tended to be higher during hibernation months than in other months but the differences were not as significant as those for calcium. Radiographic and histological analyses revealed an increase in bone resorption during hibernation. Similar to the observations reported by Doty and Nunez (1985), osteoclasts and functionally differentiated osteoblsts were absent from bone surfaces. Hibernation bone loss in the bat appeared to be mediated by osteocytes. During hibernation, confluent osteocytic lacunae were commonly found. Osteocytes were larger and the staining properties of the surrounding matrix quite different than observed during active (summer) months. Kwiecinski *et al.* (1987) concluded that hibernation bone loss (at least in the bat) was not the result of osteoclastic activity. Following arousal, endosteal surfaces become repopulated with osteoblasts and new bone formation occurred. Plasma calcium levels decreased. Osteoclasts were also found during the summer months.

Al-Badry and Taha (1983) examined water and electrolyte balance in the hibernating Oriental long-eared hedgehog. This mammal has a torpid period extending from approximately November to spring. This hibernation period is interrupted by about four to five spontaneous arousals. These investigators measured a number of plasma constituents including water content and calcium concentration in summer active, mid winter hibernating and awake (winter spontaneous arousal and spring full arousal) animals. Mean plasma water content increased by only 1.14% during hibernation compared to the summer active state. Mean plasma water content fell during arousal to a level similar to that of the summer active state. Mean plasma calcium concentration increased from 1.21 meq/l during the summer active state to 4.998 meq/l during hibernation. Mean plasma calcium concentration was 2.45 meq/l during spontaneous arousals and 1.81 meq/l during spring full arousals. Bone metabolism was not examined in these experiments but clearly bone demineralization is the most likely explanation for the hypercalcemia during the hibernating cycles.

These studies in hibernating mammals appear to indicate that the bone loss occurring during the hibernation period may proceed by a different mechanism than the bone loss that occurs in humans confined to bed for prolonged periods.

Natural animal model: American black bear (*Ursus americanus*)

Bears exhibit continuous winter dormancy for up to seven months without eating, drinking, defecating or urinating (Chapter 2 of this book; Hellgren, 1998). Black bears reduce body temperature by only a few degrees during dormancy in contrast to the "true" hibernators which lower body temperature close to that of ambient temperature. Body temperatures of hibernating bears range from 31°C to 36°C and metabolic rate averages about 50% to 70% of the basal metabolic rate of non-hibernating bears (Hellgren, 1998; Harlow *et al.*, 2004). Hibernation in small mammals involves periodic bouts of arousal which vary in frequency depending upon body size and time of winter. Do bears arouse during winter dormancy? Although observed captive bears are essentially inactive during the denning period (Donahue *et al.*, 2006a), this question has been addressed most rigorously by Harlow *et al.* (2004). These investigators measured deep core and surface skin temperature in five adult hibernating bears. In addition, ambient temperature in the vicinity of the bears' dens was also recorded. Once bears entered torpor, indicated by a drop in core body temperature about 4°C to 5°C below their active non-hibernating temperature, they remained in that state throughout the observational period without arousal to a euthermic state. The core body temperature of these bears remained stable within a 1.5°C temperature range during the entire monitoring period (December to late February). The constancy of core body temperature is strong evidence that periods of arousal do not occur in the denning bear. Daily rhythmic fluctuations in skin surface temperature were observed and the investigators interpreted these daily temperatures fluctuations as consistent with bouts of skeletal muscle contractions. Heat produced from these muscle contractions, rather than raising core body temperature and stimulating arousal was dissipated across the skin. Harlow *et al.* (2004) proposed that these episodes of muscle contractions maintain muscle strength and responsiveness in the dormant bear (see "Disuse Muscle Atrophy" section in this chapter). Pardy *et al.* (2004) have suggested that these daily bouts of muscle activity also function to provide sufficient mechanical stimuli to suppress the loss of bone mass associated with immobilization. However most likely this limited muscle activity does not provide enough mechanical stress to account for the maintenance of near normal bone mass in the hibernating bear (Donahue *et al.*, 2006a).

Floyd *et al.* (1990) studied three male captive black bears before denning, near the conclusion of winter denning and ten weeks after arousal. Denning bears were never observed outside of their dens. The bears did not shiver or have seizures except on arousal from anesthesia. No spontaneous muscular activity was observed. Blood (total) calcium concentration was not significantly different between denning and non-denning periods. In some studies, blood phosphorus levels were significantly lower during denning (Floyd *et al.*, 1990; Hellgren *et al.*, 1993), whereas in other studies blood phosphorous concentrations have shown a trend toward increasing values during hibernation (Hellgren *et al.*, 1990). The eucalcemia of the denning bear is in striking contrast to the increased blood calcium levels observed during hibernation in the bat and hedgehog and in humans during prolonged bed rest. Histological analysis of iliac crest biopsies by Floyd *et al.* (1990), revealed that trabecular bone volume did not change from summer to winter but did increase after spring arousal. Total resorption surface increased in winter and fell in spring whereas formation surface showed a modest increase in winter but then a dramatic increase in spring. Bone formation rate showed similar changes with a modest increase in winter but a significant increase in spring. Floyd *et al.* (1990) observed that mineralizing and (bone) formation surfaces were comparable in both summer and winter. This observation would be evidence that osteomalacia does not occur in the denning bear. In the denning bear, bone resorption was mediated by osteoclasts in contrast to the situation in small mammalian hibernators discussed previously. In summary, although denning bears demonstrate increased bone resorption, osteoblastic bone formation continues at a rate equivalent to that of the pre-denning period. With spring arousal, a striking increase in bone formation occurs associated with a decline in the rate of bone resorption. Interestingly, serum alkaline phosphatase levels fall during denning (Hellgren *et al.*, 1990 and 1993; David *et al.*, 1988). This observation is somewhat surprising given that alkaline phosphatase is considered a marker of bone formation (Hellgren, 1998; Gokhale *et al.*, 2001).

Donahue *et al.* (2003a and b) have used serum markers of bone metabolism to measure formation/resorption rates in bears. Type 1 collagen makes up 90% of the organic matrix of bone. During formation, the carboxy-terminal propeptide of type 1 procollagen (PICP) is cleaved off and released into the circulation. During resorption the carboxy-terminal cross-linked telopeptide of type 1 procollagen (ICTP) is released into the blood.

Serum PICP and ICTP levels have been positively correlated with histo-morphometric measurements of bone formation/resorption, respectively. In an initial study (Donahue *et al.*, 2003a) a limited number of measurements of serum PICP and ICTP levels were made in 17 wild black bears. Blood samples were collected during winter denning and active summer periods. The serum marker of bone resorption (ICTP) was significantly higher during the denning period relative to the active summer period. However, levels of the bone formation marker (PICP) were not significantly different between denning and non-denning periods. In a more recent study (Donahue *et al.*, 2003b) the time courses of serum levels of PICP and ICTP were measured in five captive black bears during pre-hibernation, hibernation and post-hibernation remobilization. During pre-hibernation, PICP levels fell but then increased early in the course of hibernation and remained elevated. Several weeks after arousal, PICP levels spiked significantly. Levels of the bone resorption marker ICTP, increased early in hibernation and remained elevated until arousal. Then ICTP levels dropped rapidly towards pre-hibernation values. The data are consistent with a picture of increased bone resorption and net bone loss during hibernation with bone formation continuing during hibernation at a rate equivalent to that of the pre-hibernation period. Upon arousal, bone formation significantly increases and bone resorption decreases thus restoring bone mass. Hence the ability of the black bear to maintain bone competence is due to the bear's unique ability to maintain bone formation during disuse and to rapidly increase bone formation during remobilization. Since in other mammals, recovery of bone lost during disuse often requires a remobilization period much longer than the original bout of disuse, this model would explain why the bear does not suffer from disuse osteoporosis despite annual periods of hibernation/disuse of five to seven months.

Donahue and co-workers have gathered additional experimental data in support of this model. The tibia was removed from each of 16 black bears of varying ages, late in their active period and the material properties (bending strength, bending modulus, porosity, ash fraction) were measured in a section of cortical bone taken from the mid-diaphysis (Harvey and Donahue, 2004). Ash fraction is a measure of bone mineral content. Although the bone samples were collected at one time point, the data are useful for assessing the cumulative effects of annual periods of disuse. The material property measurements were regressed against age of the bear. In addition, the samples were normalized by lifespan so that the bear and the human could be directly compared (since each species has a different life expectancy). Humans do not experience annual periods of disuse. For the bear, bending strength, bending modulus and ash fraction all increased with age/lifespan whereas porosity showed no change with lifespan/age. In the human, bending strength, bending modulus and ash fraction also increased with age/lifespan as in the bear. However, unlike the bear, in humans bone porosity increased with age/lifespan. The data indicate that the mechanical integrity of bear cortical bone are not compromised with age despite annual periods of disuse equivalent in duration to active periods. In humans, unlike the bear, bone cortical porosity increased with age even though humans have no regular cycle of disuse/active periods. In a more recent publication, Harvey et al. (2005) measured the tensile strength, elastic modulus (both derived from a linear applied force to the bone sample) and porosity and ash content of cortical tibial bone samples from 27 black bears of various ages. The samples were obtained from bears late in their active period. Tensile strength, elastic modulus and porosity did not display any relationship to age of the bear. Ash fraction increased with age. The tensile strength for the bear was comparable to the value found in other species including the human. These observations in the bear are remarkable considering that this animal is inactive for about six months each year.

In an allied study, Pardy et al. (2004) examined bone mass and bone architecture in 22 pre-denning (autumn) and 23 post-denning (spring) black bears. The following measurements were made; bone mineral density, bone mineral content, trabecular bone volume and trabecular bone micro-architecture. No statistically significant differences were noted between pre-and post-denning bears in any of the measured parameters. Maintenance of mineral status (calcium) is particularly noteworthy when considering that denning bears are eucalcemic (measured as total calcium) and anuric (Floyd et al., 1990). Equally remarkable is the maintenance of trabecular bone volume and micro-architecture during the denning period. One would have expected the lack of mechanical stimuli during hibernation to lead at least to an alteration of trabecular micro-architecture between the pre and post-denning periods. Donahue et al. (2006a) have summarized available data on bone homeostasis in the hibernating bear. They consider two possible interpretations of the current observations. (1) Bone is not lost to disuse during hibernation because bone formation and bone resorption remain coupled even though the overall rate of bone turnover has increased. (2) Some bone is lost to disuse during hibernation because resorption exceeds formation but the lost bone is rapidly recovered during remobilization which occurs shortly after spring arousal. In either case, bone homeostasis in the bear during hibernation clearly differs from that occurring in other immobilized mammals in that the bear experiences very little disuse osteoporosis.

In a more recent study, Donahue *et al.* (2006b) measured seasonal changes in the serum levels of ionized calcium, parathyroid hormone (PTH), 25 (OH) vitamin D and leptin in 16 different female black bears. Blood sample collection times encompassed an active pre-hibernation period, hibernation and an active post-hibernation remobilization phase. The mean ionized calcium concentration was significantly higher in the hibernation and post-hibernation periods compared to pre-hibernation. Mean PTH concentration was highest post-hibernation compared to pre-hibernation and hibernation. PTH concentration did increase during hibernation relative to pre-hibernation but the increase was not statistically significant. Mean 25 (OH) vitamin D levels did not show any seasonal variation while mean leptin levels did not change between pre-hibernation and hibernation but did significantly decrease in the post-hibernation period. The observation that PTH levels increased (relative to pre-hibernation) during hibernation and the post-hibernation phases is puzzling given that the ionized calcium concentration was increased during these same two periods (compared to pre-hibernation). The serum ionized calcium concentration is a sensitive regulator of PTH levels so the data of Donahue *et al.* (2006b) begs the question as to what is the stimulus responsible for the elevated PTH concentration in the bear. This question is addressed in the next section.

Bone homeostasis in the bear: research questions

The denning bear is able to maintain a near normal bone mass despite many months of inactivity. However, the limited daily bouts of skeletal muscle activity described by Harlow *et al.* (2004) are unlikely to provide sufficient mechanical load to account for this maintenance of bone mass. In addition, during hibernation the bear remains eucalcemic despite being anuric. By contrast, humans immobilized for prolonged times show increased bone resorption (uncoupled from formation), calcium release from bone, hypercalcemia, hypercalcuria and secondarily suppressed levels of PTH and calcitriol (Zerwekh *et al.*, 1998; Clouston and Lloyd, 1987). The data from other hibernating mammals indicate that progressive bone loss occurs during hibernation followed by an abrupt renewal of bone formation upon spring arousal. Furthermore, the physiological state of the denning bear is complicated by the concurrence of functional renal failure. During hibernation, glomerular filtration rate (GFR) decreases by about 60% (Brown *et al.*, 1971) and the small amount of urine formed is completely reabsorbed by the urinary bladder (Nelson *et al.*, 1975). In humans, secondary hyperparathyroidism with elevated levels of PTH develops early in the course of kidney failure at a time when GFR has fallen by only 30% below normal values (Rodriguez *et al.*, 2005). There are a number of factors leading to secondary hyperparathyroidism (Martin *et al.*, 2004). However, the key factor appears to be the retention of phosphate (phosphorus) as a result of the decrease in GFR in chronic renal failure. Phosphate (phosphorus) retention leads to a number of effects: hypocalcemia, depressed renal production of calcitriol, and parathyroid gland hyperplasia and increased PTH secretion. Since calcitriol is a negative regulator of PTH secretion and parathyroid cell proliferation, low levels of calcitriol will also result in elevated PTH levels. What is fascinating is that the histological changes in bone induced by parathyroid hormone share some of the features described in the bone of the hibernating bear. In secondary hyperparathyroidism, there is an increased rate of bone formation, increased bone resorption, extensive osteoclastic and osteoblastic activity and a progressive increase in endosteal peritrabecular fibrosis (Martin *et al.*, 2004). In mild to moderate primary hyperparathyroidism, the following bone histological changes have been described (Fitzpatrick and Heath, 2001); reduced cortical thickness, preserved or even increased trabecular bone volume, and increased bone turnover. There are signs of increased osteoclastic bone resorption as well as evidence of increased bone formation (increased osteoid surface with normal mineralization, increased osteoblast numbers and increased mineral apposition rate). The anabolic actions of PTH on bone have been reviewed recently by Hock (2001). Studies in the rat have shown that PTH can increase trabecular and cortical bone mass by stimulating bone formation. These anabolic effects are observed with intermittent administration of PTH whereas catabolic effects (bone resorption) seem to predominate when PTH is given continuously. In human studies the initial biochemical and histomorphometric changes to PTH administration are consistent with an immediate stimulation of bone formation and a later increase in resorption. In rats, increases in bone matrix proteins and bone forming surfaces in trabecular bone occur within 24 hours of the first injection of PTH while resorption measures remain unchanged. During induction of the anabolic response, PTH regulates many osteoblastic genes as discussed by Hock (2001). The osteoblast and its progenitor cell are a primary target of PTH.

Does the hibernating bear develop (secondary) hyperparathyroidism? The bear has a significant decrease in GFR and the reduced volume of produced urine is completely reabsorbed by the bladder. However reported changes in blood phosphate (phosphorus) levels have not been consistent. Investigators have described both high and low values compared to the pre-denning state (Floyd *et al.*, 1990; Hellgren *et al.*, 1990 and 1993). According to the measurements of Donahue *et al.* (2006b) the bear during hibernation and in the post-hibernation remobilization phase does have a hyperparathyroid state. Compared to pre-hibernation, PTH levels are definitely higher in the post-hibernation period and in addition during hibernation, PTH levels are also high given the fact that PTH secretion should be suppressed by the elevated ionized calcium concentration. An important related question would be whether the hibernating bear has modified this state such that intermittent or pulsatile elevations of PTH occur, rather than a continuous elevation given that experimentally the anabolic effects are observed with intermittent PTH injections rather than a continuous infusion. My speculation is that PTH is an important signal molecule accounting at least in part, for the increased rates of bone formation and resorption in this animal, and Donahue *et al.* (2006b) also suggest that PTH probably plays a dominant role in preserving bone formation during the hibernation and post-hibernation (remobilization) periods. As already noted, the stimulus for this hyperparathyroid state is unclear. One possibility, although purely speculative, is that the hibernating bear does develop secondary hyperparathyroidism which then "progresses" into a tertiary or autonomous state.

Perhaps, a more fundamental question is whether the denning bear utilizes mechanisms present in all mammals to maintain bone mass and homeostasis or produces a unique osteoregulatory substance as suggested by Floyd *et al.* (1990). There have been attempts to detect unique molecules involved in bone homeostasis in the blood of the bear. Milbury *et al.* (1998) using high performance liquid chromatography analyzed the plasma from black bears across four seasons and from three areas. The results were compared with the database generated from the National Aeronautics and Space Administration (NASA) human bed rest study. This study showed a clear difference in the metabolite profiles between the plasma of ambulatory and immobile subjects. Milbury *et al.* (1998) highlighted six chemically uncharacterized compounds which changed significantly but differentially in wintering bears and immobile humans. Unfortunately the nature of these low molecular weight compounds has yet to be clarified. Overstreet *et al.* (2003 and 2004), have examined the effects of bear serum on the properties of human osteoblast cultures. In the first study (Overstreet *et al.*, 2003) two serum samples were obtained from each of four different bears. One of the samples was acquired during hibernation and the other while the bears were free ranging. Human osteoblasts were incubated for 24 hours in medium containing 10% bear serum or 10% calf serum or no serum. The end points measured were cell proliferation and phenotype expression (osteocalcin and type 1 collagen). Four of the eight bear serum samples induced significantly higher proliferation rates in osteoblasts than did medium containing fetal calf serum or no serum. Levels of mitogenic effect by different bear sources varied with the source (individual bear) and the specific time (season) the sample was taken. Osteoblasts incubated in bear serum also displayed higher mRNA levels for phenotype markers osteocalcin and type 1 collagen. The molecular weight of the mitogenic factor(s) was found to be greater than 50 kDa. Overstreet *et al.* (2004) also observed that osteoblasts incubated in bear serum for three days detached from the culture plates although no differences were observed between serum samples obtained from free ranging and hibernating bears. Cells incubated in medium containing calf serum or no serum remained adherent. Analysis of mRNA showed elevated integrin transcript levels in osteoblasts cultured in bear serum compared with cultures in fetal calf serum or no serum. Integrins are adhesion receptors believed to mediate cellular sensing of mechanical loads. The studies by Milbury *et al.* (1998) and Overstreet *et al.* (2003 and 2004) suggest that bear serum may contain unique factors with bone homeostasis regulatory properties. However, clearly more research needs to be done to clarify the nature of these molecules.

In summary, hibernating black bears maintain a normal (or near normal) bone mass and bone architecture by having a bone formation rate at least equal to that of the pre-hibernation period. Since the denning bear is inactive, bone formation is not initiated primarily by mechanical forces and hence must be driven by metabolic signal molecules. Given the current published data available, it is probably reasonable to assume as a first approximation that the bear relies on metabolic signal molecules present in all mammals but modified in some way. Whether the bear has evolved unique molecules for regulating bone homeostasis has not been clearly demonstrated in the published studies. As noted at the beginning of this chapter, the adaptations by which the denning bear minimizes disuse osteoporosis and disuse muscle atrophy (discussed in the next section) are clearly only a component of a whole set of adaptations which also enables the denning bear to avoid the nitrogen/protein metabolic consequences of renal failure with anuria. A future challenge will be not only to understand the adaptations relevant to bone, muscle and renal physiology but to also understand the inter-relationships of these adaptations as part of a larger integrated whole.

Although it is clear that the bear has solved the problem of disuse osteoporosis, it is equally clear that we do not understand how the bear accomplishes this biological feat. What follows are but a few suggested research questions.

(1) There are two published studies containing blood calcium measurements in the bear (Floyd *et al.*, 1990; Donahue *et al.*, 2006b). Floyd *et al.* (1990) measured total calcium concentrations and reported that the hibernating bear remained eucalcemic. Donahue *et al.* (2006b) reported that ionized calcium concentrations were increased during hibernation and the early post-hibernation period. Blood calcium exists in three forms: complexed with organic anions, protein bound and ionized. The ionized calcium is the biologically active fraction. Could the eucalcemia noted in the denning bear by Floyd *et al.* (1990) be due to changes in calcium binding between the pre- and denning states? David *et al.* (1988) in their abstract reported that serum calcium levels were highest in the denning period although still within the normal range for humans. However actual values are not given. In the human, the concentration of calcium in the extracellular fluid represents a dynamic balance between intestinal absorption, renal excretion and bone formation/resorption (Friedman, 2000). In the denning bear, intestinal absorption and renal excretion do not play a role in determining the extracellular calcium concentration since this animal has no intake and is anuric. Hence during denning, the extracellular calcium concentration will reflect the balance between the rates of bone formation and resorption. If the rate of resorption exceeds that of formation the bear should become hypercalcemic. Conversely hypocalcemia would result if the rate of bone formation is greater than that of resorption. If the denning bear is eucalcemic then the rates of bone formation and resorption must remain coupled and essentially equivalent. If the denning bear is hypercalcemic then the rate of bone resorption must exceed the rate of formation. This type of analysis however, requires reliable serial measurements of ionized blood calcium concentrations in the denning and non-denning periods. The observation of Donahue *et al.* (2006b) that the denning bear is hypercalcemic compared to the pre-hibernation bear needs to be confirmed.

(2) There are many known bone homeostasis regulatory compounds of which three of the most important are PTH, calcitriol and leptin. The data of Donahue *et al.* (2006b) indicate that of these three regulatory factors PTH appears to play a significant role in maintaining bone mass during the denning period although Donahue *et al.* measured 25 (OH) vitamin D not calcitriol. However, the stimulus for this hyperparathyroid state in the hibernating bear is unknown. The elevated levels of ionized calcium in the hibernating bear should result in a decrease in PTH secretion. What then are the signals leading to a hyperparathyroid state in the hibernating bear? The answer to this question is currently unknown. As already discussed, the effects of PTH on bone (anabolic versus catabolic) for example, appear to be dependent, at least experimentally, upon whether PTH is administered intermittently or continuously. Hence, measurements of blood levels of these three molecules should be made not only in the pre- and denning periods but also serially during the day to record any peaks and valleys in blood concentrations.

(3) The published studies using bears appear to be based on the implicit assumption that bone cells (osteoblasts and osteoclasts) in the bear are similar to those of other mammals, but that the bear differs in terms of the nature of its regulatory factors. It would be important to establish whether bear bone cells are similar to those of other mammals and this question could conceivably be answered by studying bear osteoblasts and osteoclasts in culture. Do bear bone cells display similar genetic and phenotypic properties as well as similar responses to known regulatory factors as the bone cells of other mammals?

The hibernating bear does indeed appear to be a metabolic marvel and in the next section we will explore how the bear minimizes disuse muscle atrophy during its prolonged denning period.

Disuse Muscle Atrophy

Maintenance of muscle mass

Muscle atrophy occurs during prolonged periods of reduced muscle activity such as bed rest, limb immobilization or space flight. The atrophy occurs due to both a decrease in muscle protein synthesis and an increase in the rate of proteolysis (Powers *et al.*, 2005). Several studies have addressed the time line for these two processes.

Tucker *et al.* (1981) measured the rate of protein synthesis in the gastocnemius muscle of a group of adult rats subjected to immobilization of both hindlimbs for seven days. A group of rats not immobilized served as a control group. In the immobilized rats, the gasrocnemius muscle was positioned at a muscle length that was less than its resting length. After one week of immobilization, the protein content of the gastrocnemius (a predominantly fast-twitch muscle) had decreased by 27% compared to control values. Over the first six hours of immobilization, the fractional rate of protein synthesis fell from a control rate of 6.8%/day to 3.6%/day. Thereafter the fractional rate of protein synthesis remained unchanged (at a value of about 3.5%/day) for the duration of the immobilization period. Tucker *et al.* (1981) calculated that the difference between the fractional rates of protein synthesis in the immobilized and control gastrocnemius muscle could completely account for the 27% decrease in protein content after seven days of immobilization. However there are several problems with this calculation. One problem in particular is that the investigators assumed that degradation rate remained at the control level throughout the seven-day period. Booth (1982) also reported a decrease of approximately 35% in the rate of protein synthesis in rat skeletal muscle during the first six hours of limb immobilization. Booth (1982) proposed that decreased protein synthesis is the initial change in protein turnover that results in the net loss of muscle protein.

Goldspink (1977) measured rates of protein synthesis and protein breakdown in the soleus (predominantly slow-twitch) and extensor digitorum longus (predominantly fast-twitch) muscles of growing rats during seven days of unilateral hindlimb immobilization. The muscles of the contralateral limb served as internal controls. The data considered here are for experiments in which the muscles were fixed in a shortened position. In the soleus muscle, a decreased rate of protein synthesis was observed as early as six hours of immobilization consistent with the data of Tucker *et al.* (1981) and Booth (1982). The average rate of protein synthesis (per mg muscle) declined by about 65% compared to control by day 2 of immobilization and remained unchanged at that value for the next five days of the experiment. Changes in the rate of protein breakdown were slower in onset than the changes in synthetic rate. The rate of protein breakdown (per mg muscle) increased by about 30% compared to control by day 2 and remained relatively stable for the duration of immobilization. Rates of protein synthesis and breakdown measured in the extensor digitorum longus muscle demonstrated similar trends but of a lesser magnitude than those of the soleus.

Thomason *et al.* (1989) studied protein metabolism in the skeletal muscle of rats randomly assigned to one of three groups: control, five hours of hindlimb suspension and seven days of hindlimb suspension. Synthesis rates of total mixed protein and myofibril protein were measured in the soleus, plantaris, medial gastrocnemius, adductor longus and quadriceps muscles. The soleus and adductor longus are predominantly slow-twitch tonic muscles while the other three are predominantly fast-twitch phasic muscles. In the tonic muscles, there was a decline in fractional rates of protein synthesis (total mixed and myofibril) evident within five hours of hindlimb suspension and a significant decrease in these rates by seven days. In the phasic muscles, fractional rates of protein synthesis actually increased at five hours but these rates were significantly decreased below control values at seven days of hindlimb suspension. The investigators modeled the decrease in soleus muscle myofibril protein fractional synthesis rate as a first order process with a half time of 0.3 days reaching a new steady state value equal to the seven-day synthesis rate. They also calculated a continuous function for the rate of change of soleus muscle myofibril protein content during hindlimb suspension from previously published data on the time course of soleus muscle myofibril protein content decrease with hindlimb suspension. Myofibril protein content rapidly declined by about 70% over 14 days reaching a new steady state value by day 21 of hindlimb suspension. From these two functions (myofibril protein synthesis rate and the rate of change of myofibril protein content), they derived a function for the time course of soleus muscle myofibril protein degradation rate over an extended period of 28 days of hindlimb suspension. Their data and calculations showed the following picture. Fractional rate of protein synthesis decreased rapidly over the first two to three days from a baseline of 5.9%/day to a new steady state value of 2.4%/day. The fractional rate of protein degradation began to increase by day 2–3 reaching a maximum of about 15%/day (from a baseline of 5.9%/day) by day 15. Thereafter the fractional rate of protein degradation decreased to a new lower steady state value of 2.4%/day by day 24 of hindlimb suspension. These changes in the rates of synthesis and degradation result in a 75% decrease in soleus muscle myofibril protein content by about day 21. Myofibril protein content remains stable thereafter. The data of Thomason *et al.* (1989) clearly show that although a decrease in protein synthetic rate is the initial event, the genesis of disuse muscle atrophy, at least in the rat hindlimb suspension model, involves an increase in protein degradation rate as well as a decrease in rate of protein synthesis. Thomason *et al.* (1989) also measured soleus muscle beta-myosin heavy chain mRNA concentration as a fraction of control values after five hours and seven days of hindlimb suspension. There was no decrease in beta-myosin heavy chain mRNA concentration after seven days as would have been expected if a transcriptional control mechanism was driving changes in translation. This data are consistent with regulation of mRNA translation by ribosomal machinery as a mechanism for the suppression of myofibril protein synthesis. As discussed in Chapter 2, the key to translational suppression is reversible phosphorylation control over the activities of ribosomal initiation and elongation factors.

Booth (1977) has examined the time course of muscle atrophy in rats with both hindlimbs immobilized. Only data for muscles fixed at a position slightly less than resting length will be considered. Actual weight losses for the gastrocnemius and soleus muscles were 45% and 52%, respectively over the first 21 days with no further significant weight loss between day 21 and day 35, the last observation point. Using the difference in values between days 0, 3, 6, 10 of immobilization and the apparent steady state value on day 28, the rates of weight loss for the gastocnemius, soleus and quadriceps were exponential between the second and tenth day of immobilization. The initial time lag of about two days before the rate of weight loss becomes exponential is consistent with the data of Goldspink (1977) and Thomason *et al.* (1989) demonstrating a two- to three-day delay in the onset of changes in the rate of protein breakdown after initiation of hindlimb immobilization or suspension. In addition, Booth (1977) reported an exponential loss in the total muscle tissue activity of citrate synthase and the total amount of cytochrome c in the gastrocnemius muscle over days 2 to 10 using the differences between values obtained on days 0, 2, 4, 7 and 10 and steady state values on day 28 of immobilization. However, cytochrome c content and citrate synthase total activity decreased faster than gastrocnemius weight. Mass specific cytochrome c concentration and mass specific citrate synthase activity fell but then rose after day 7 of immobilization since these two proteins were approaching their new steady state values while muscle weight was still decreasing. Hence changes in mass specific citrate synthase activity do not necessarily reflect what is happening to total tissue activity.

In summary, the data obtained from animal experiments using hindlimb suspension or immobilization give the following general model for the genesis of muscle atrophy. The initial event is a rapid decline in the rate of protein synthesis with a new steady state developing rather quickly. A change in the rate of protein degradation occurs subsequently with an initial increase in the rate followed by a decrease to a new lower steady state value equal to the steady state value achieved for protein synthesis. The integration of these changes in the rates of synthesis and degradation results in losses of muscle weight, total muscle protein content and levels of mitochondrial oxidative enzymes (such as citrate synthase) which are exponential. The experimental animal used by Tucker *et al.* (1981), Goldspink (1977), Thomason *et al.* (1989) and Booth (1977) was the rat so that the timeline for changes in the rates of synthesis and degradation and establishment of new steady state values are probably specific for this animal. However, it is probably reasonable to consider this model generalizable.

Protein degradation: mechanisms

Powers *et al.* (2005) and Reid (2005) have recently reviewed mechanisms involved in muscle protein degradation. A number of proteolytic systems contribute to the breakdown of skeletal muscle: lysosomal proteases, calcium activated proteases (i.e. calpain) and the proteasome system. The calpain and proteasome systems are the most important. The bulk of muscle proteins exist in actomyosin complexes. The proteasome system can degrade monomeric contractile proteins (i.e. myosin and actin) but not intact actomyosin complexes. Hence, myofilaments must be released from the sarcomere as monomeric proteins before degradation by the proteasome pathway. Calpains are calcium dependent cysteine proteases that are activated in skeletal muscle during period of disuse. Calpains bring about dissociation of actomyosin complexes by cleaving cytoskeletal proteins that anchor the contractile elements. Hence calpain activation would be a first step. The calpain mediated release of myofilaments (actin and myosin) would allow degradation of these proteins by the proteasome pathway. In this pathway, proteins can be degraded by either the 20S core proteasome or the 26S proteasome (composed of the 20S core with a regulatory 19S complex connected to each end). In the 26S proteasome pathway, the poypeptide ubiquitin covalently binds to protein substrates and marks them for degradation by the 26S proteasome, a multicatalytic enzyme complex. However, the 20S core proteasome can selectively degrade oxidatively modified proteins without requiring prior binding by ubiquitin. Markers of proteasome-ubiquitin pathway activity are increased by conditions that diminish muscle use, including limb immobilization, gravitational unload-ing and denervation. Also oxidative stress (defined as occurring when the production of reactive oxygen species (ROS) exceeds the rate of removal) is believed to be an important link between muscle disuse and increased muscle proteolysis.

ROS production in muscles is increased during periods of disuse (Powers *et al.*, 2005). Examples of ROS include superoxide, peroxyl radi-cals, hydroxyl radicals and hydrogen peroxide. ROS can react with lipids, proteins and nucleic acids to alter their structure and function. Powers *et al.* (2005) have summarized the evidence that ROS accumulate in inactive muscles and that this increased level of ROS contributes to disuse muscle atrophy by increasing the rate of proteolysis via the proteasome system and probably also by inhibiting the rate of protein synthesis. Hudson and Franklin (2002) have proposed that a key pathogenic feature of disuse mus-cle atrophy is the relative balance between levels of ROS and antioxidants which determines the occurrence and severity of the muscle atrophy. How-ever, as noted by Reid (2005) ROS are just one of a number of factors linking altered muscle use to activity of the ubiquitin proteasome pathway [see Fig. 2 in review of Reid, (2005)].

Animal models

As already alluded to in the previous sections, several experimental ani-mal models have been developed for studying the pathophysiology of dis-use atrophy: hindlimb immobilization (HI) and hindlimb suspension (HS) (Fitts *et al.*, 1986; Booth, 1982; Thomason and Booth, 1990). The two models produce similar but not completely equivalent effects. Both models result in muscle atrophy. For example, 14 days of HS and HI lead to 50% and 40% decreases, respectively in the weight of the soleus muscle (predominantly slow-twitch). Fourteen days of HS produces a 15% decrease in the weight of the extensor digitorum longus (predominantly fast-twitch). Both models affect preferentially slow-twitch (type 1) fibers with the result that there is a change in the relative proportions of type 1 and type 2 (fast-twitch) fibers. For example, the rat soleus muscle is composed of approximately 70%–90% type 1 and 10%–30% type 2 fibers, but after four weeks of HS there are about equivalent numbers of type 1 and 2 fibers (Thomason and Booth, 1990; Templeton *et al.*, 1984; Templeton *et al.*, 1988). The cause of this shift in relative distribution of fibers has been addressed in several studies. Booth (1982) reported an absolute reduction in the number of slow-twitch (type 1) but no significant change in the absolute number of fast-twitch (type 2) fibers in the cross-section of the soleus muscle from limbs immobilized for four weeks. The result was a reduction in the percentage of slow-twitch fibers after limb immobilization as well as a reduction in the total number of (type 1 and 2) fibers. Nicks *et al.* (1989) found that right forelimb immobilization in rats for eight weeks did not lead to any change in mean total fiber number (type 1 and 2 combined) in the triceps brachii muscle compared to the same muscle in the contralateral (non-immobilized) control limb. Templeton *et al.* (1988) noted that in rats subjected to four weeks of HS, there was no change in mean total fiber number in the soleus muscles compared to previously excised control soleus muscles from the same rats prior to HS. The consensus appears to be that total fiber number is not changed by HI or HS. As discussed by Templeton *et al.* (1988), the mechanism most likely responsible for the change in fiber composition of the soleus (with HS) is transformation of type 1 fibers into type 2 fibers. Total fiber number remains the same because the decrease in number of type 1 fibers is matched by the increase in number of type 2 fibers. Both HS and HI also result in a decrease in fiber cross-sectional area (CSA). In the study of Nicks *et al.* (see above) mean fiber CSA in the triceps brachii of the immobilized limb decreased by 42.1% compared to the same muscle in the contralateral control limb after eight weeks of immobilization. This mean decrease of 42.1% in fiber CSA closely approximated the 38% mean decrease in muscle wet weight. In the rodent (rat, mouse and hamster), HS results in a reduction in mean CSA of both type 1 and type 2 fibers in the soleus muscle of between 15%–60% (Thomason and Booth, 1990). In the study of Templeton *et al.* (1988), four weeks of HS resulted in a 41% reduction in total fiber (type 1 and 2) CSA of the rat soleus muscle. This change largely accounted for the 47% decrease in soleus mean wet weight. As reviewed by Thomason and Booth (1990), the soleus muscle in rodents subjected to HS shows most of the decrease in muscle fiber CSA and most of the change in the relative proportions of type 1 and type 2 fibers in the first two to four weeks of HS even though some studies included prolonged periods of suspension up to 206 days. In summary, the consensus is that the muscle atrophy (decrease in muscle mass) induced by HS or HI is due to a reduction in type 1 and type 2 fiber size with no change in total fiber number. The change in relative proportions of type 1 and type 2 fibers is due to a transformation of type 1 to type 2 fibers.

In general, the effects of HS and HI on the contractile properties of muscle will reflect the structural changes discussed in the previous paragraph, i.e. a loss of contractile proteins (decrease in fiber size) and an alteration in the relative proportions of type 1 and type 2 fibers with no change in the number of total fibers. HS and HI have somewhat different effects on muscle function since these two techniques do not necessarily lead to equivalent alterations in these structural properties. Both HS and HI reduced the isometric twitch duration in the soleus (Fitts *et al.*, 1986). However the suspension mediated effect on twitch duration was entirely restricted to an altered contraction time. In contrast, HI reduced both contraction time and relaxation time. Two weeks of hindlimb suspension in rats produced a two fold increase in the maximal shortening velocity of the (slow-twitch) soleus muscle, whereas a comparable period of hindlimb immobilization did not significantly change the maximal shortening velocity of the soleus muscle. However, after six weeks of HI maximal shortening velocity did increase but only by 20% or less (Fitts *et al.*, 1986). The lack of a large change in maximal shortening velocity with HI is surprising since HI does result in a reduction in the percentage of slow-twitch fibers as already noted. Both models reduced peak tetanic tension (a measure of muscle strength) and these reductions were comparable to the reductions observed in muscle mass. The decrease in peak tetanic tension indicates a greater atrophic effect on myofibrillar protein and hence the number of force generating sites (cross bridges) compared with non-contractile protein. As already noted, both HS and HI primarily affect type 1 muscle fibers and thus slow-twitch muscles. In keeping with this observation, HS had no effect on the contractile properties of the fast-twitch extensor digitorum muscle except for a decrease in peak tetanic tension. Two weeks of HS caused a 79% reduction in peak tetanic tension in the soleus but only an 11% reduction in the extensor digitorum longus. HI had a somewhat greater effect in the extensor digitorum longus, causing prolonged twitch duration as well as a reduced peak tetanic tension. Fitts *et al.* (1986) also point out another difference between these two models. Both produce decreased muscle activity but HS completely eliminates weight-bearing such that any contractile activity is unloaded and characterized by low forces. With HI, the aggregate electromyogram (number of spikes per unit time) is reduced but isometric contractions with a high force component may still occur. Based on this comparison, HS may be a better model for the effects of the weighlessness of space travel and HI may be a better model for bed rest.

Based largely on these two models (HI and HS), a number of factors have been described which influence the extent of muscle atrophy during inactivity/immobilization (Hudson and Franklin, 2002; Booth, 1982). Muscles immobilized in the shortened position suffer significantly greater atrophy than muscles fixed in a stretched position. In fact, fixation of muscles in a lengthened position can result in delay of onset of muscle atrophy or result in actual muscle enlargement (Goldspink, 1977; Booth, 1977). Fiber type, as already discussed, is an important variable. Muscle atrophy is greater in slow- twitch compared to fast-twitch fibers. The age of the animal also seems to influence the rate of muscle atrophy in that atrophy decreases with age. There are also differences among animal species. Hudson and Franklin (2002) reviewed limb immobilization studies in a number of different mammalian species including humans, and when the data are normalized to the same time period of immobilization there was a significantly greater rate of muscle atrophy in smaller compared to larger mammals. For example, when normalized to an immobilization period of 12 days, the rate of atrophy varied from 9% in humans to 45% in mice. However, given the observation that the rate of muscle weight loss (atrophy) is exponential, Hudson and Franklin (2002) would have had to know the half-times for each mammal to correctly normalize rates of atrophy to a given end point such as 12 days.

Studies in humans and small mammal hibernators

Humans

Berg *et al.*, (1997) measured muscle function, size and fiber composition in seven healthy men (mean age 28 years) subjected to six weeks of head-tilt down bed rest. Knee extensor torque (a measure of muscle strength) and electromyographic activity were measured the day before and two days post-bed rest. Muscle biopsies of the vastus lateralis were performed before and on day 37 of bed rest. Magnetic resonance imaging tomography was used to measure knee extensor (quadriceps femoris) cross-sectional area (CSA) before and on day 37 of bed rest. The CSA was used as a measure of muscle size. Mean maximum voluntary isometric and concentric torque (i.e. voluntary muscle strength) decreased by 25%–30% after bed rest. Mean quadriceps femoris CSA decreased by 13.8% after 37 days of bed rest. Force-velocity characteristics of the knee extensor muscle group did not show any alteration after bed rest. There was no change in maximal shortening velocity. The proportions and sizes of muscle fiber types showed the following changes. The mean CSA of type 1 fibers decreased by 18.2%. Type 2 fibers also showed a reduction in mean CSA but it was not statistically significant. Assuming that bed rest did not change total fiber number, then using the data from Table 1 of Berg *et al.* (1997), one can calculate total fiber (type 1 and 2 combined) CSA. Six weeks of bed rest resulted in a 16.3% decrease in total fiber CSA which value agrees closely with the 13.8% decrease in the quadriceps femoris measured by MRI tomography. Muscle fiber relative percentages averaged 52.6% and 47.1% for types 1 and 2, respectively before bed rest and were not significantly altered after bed rest. Myosin heavy chain isoforms were examined and the relative proportions of type 1 and 2 myosin heavy chain were not changed by bed rest consistent with the lack of effect of bed rest on fiber type relative proportions. How do these results in humans compare to the animal models (hindlimb suspension and immobilization) discussed in a previous section?

In rats, the weight loss for the quadriceps was exponential between the second and tenth day of immobilization. The calculated half-time was 4.8 days (Booth, 1977). Assuming that a similar half-time applies to decreases in the CSA of the quadriceps in the study of Berg *et al.* (1997) in humans, an apparent steady state would be achieved between days 20 and 25 of bed rest. Using this half-time, the time course for the decrease in quadriceps femoris CSA would be as follows. Half of the decrease in CSA (about 7%) would occur by day 5 of bed rest and by days 20 and 25, the decrease in CSA would be 13.1% and 13.6%, respectively. If this assumption is correct, then longer periods of bed rest beyond 25 days would not result in any further muscle atrophy. The decrease in quadriceps CSA of about 14% measured at day 37 of bed rest would have been achieved by day 25. In the animal models muscle force-velocity characteristics are altered; hindlimb suspension produced a two fold increase in maximal shortening velocity and prolonged hindlimb immobilization a 20% increase. Force-velocity characteristics were not changed by six weeks of bed rest in humans. In the animal models, the decreases in muscle strength and muscle mass are comparable, whereas in the study of Berg *et al.* (1997) in humans, muscle strength decreased by 25%–30% but muscle size only decreased by 14%. Hence in humans, the greater loss of strength relative to reduction in muscle size suggests other factors may be involved such as reduced neural input. In the animal models, type 1 fibers are preferentially affected resulting in a change in the relative proportions of type 1 and type 2 fibers as discussed in the section on animal models. It is interesting that in the study by Berg *et al.* (1997) in humans, type 1 fibers were preferentially affected in that type 1 but not type 2 fibers showed a significant decrease in mean cross-sectional area with bed rest. However, there was no change in the relative proportions of type 1 and type 2 fibers in humans after six weeks of bed rest. Hence, there are significant differences between the effects of hindlimb suspension and immobilization in animals and the effects of bed rest in humans.

Small mammal hibernators

As discussed in Chapter 2, small mammal hibernators display periodic arousals so that the hibernation season actually consists of a series of hibernation bouts. In the studies described in this section, only the one by Yacoe (1983) examined muscle alterations in different stages of the hibernation bout.

Wickler *et al.* (1991) and Steffen *et al.* (1991) studied disuse muscle atrophy in hibernating golden-mantled ground squirrels. Wickler *et al.* compared a group of squirrels after six months of hibernation with a matched active pre-hibernator group. Steffen *et al.* compared three groups: summer active squirrels, squirrels that had been hibernating for six months and winter "active" squirrels (maintained at an ambient temperature of 23°C for the same six-month period as the hibernating group). However, Steffen *et al.* observed that the winter active squirrels were actually quite inactive. They assumed a similar posture as the hibernating group, were very lethargic in their behavior and primarily engaged in sleep. Body temperatures were not determined and so it is not known whether these squirrels maintained normothermia or became somewhat hypothermic. Winter active and hibernating squirrels showed comparable reductions in mean muscle weight (15%–20% for the soleus, plantaris and extensor digitorum longus muscles) compared to the summer active group. Wickler *et al.* reported that squirrels that had hibernated for six months demonstrated significant weight loss of the gastrocnemius (14% decrease), semitendinosus (42% decrease) and soleus (23% decrease) compared to active squirrels prior to hibernation. These changes in muscle mass were not due to loss of water since mass specific protein concentration did not change during hibernation in the gastrocnemius, semitendinosus or soleus muscles. Steffen *et al.* measured a significant decrease in total protein content of the extensor digitorum longus but not the soleus muscle after hibernation. Steffen *et al.* also measured cross-sectional areas (CSA) of slow-twitch fibers in the soleus and fast-twitch fibers in the extensor digitorum longus. In both hibernating and winter active squirrels, the mean CSAs of slow- and fast-twitch fibers were reduced by 17% to 31% compared to the summer active group. These reductions in fiber CSA approximated the decreases in muscle mass. Absolute mean DNA contents in the soleus and extensor digitorum longus muscles were not significantly changed in hibernating or winter active squirrels compared to the summer active group indicating that the muscle atrophy was due to a reduction in cellular dimensions rather than a loss of cells. In both studies (Steffen *et al.* and Wickler *et al.*) mass specific citrate synthase activity in muscle tissue was increased in hibernating squirrels compared to the active control group. Steffen *et al.* did not measure citrate synthase activity in winter active squirrels. However, the picture for total citrate synthase activity (mass specific activity × muscle weight) was less clear. Compared to the active control group, hibernating squirrels had higher total muscle tissue citrate synthase activity in the gasrtrocnemius, lower total muscle tissue activity in the semitendinosus and about the same total muscle tissue activity in the plantaris. In contrast, total muscle tissue citrate synthase activity in the rat was decreased in muscles undergoing immobilization induced atrophy (Booth, 1977). Wickler *et al.* (1987) examined muscle function in hibernating hamsters after approximately 24 days of hibernation compared to a non-hibernating control group. In general, the results were similar to those obtained in the golden-mantled ground squirrel. Compared to the control group, the hibernating hamsters had significant decreases in the mean weight of the semitendinosus (63% reduction) and gastrocnemius/plantaris (30% reduction) muscles. Compared to the control group, the hibernating hamsters had an increase in the mass specific citrate synthase tissue activity in both muscle groups, but total muscle tissue citrate synthase activity (mass specific activity × muscle weight) was lower in the hibernating group than in the non-hibernating hamsters.

The study by Yacoe (1983) gives important insights into the dynamics of protein metabolism in a hibernating small mammal. Yacoe measured *in vivo* rates of fractional protein synthesis (as incorporation of radioactively labeled phenylalanine into protein) in the pectoralis muscle and liver of the big brown bat during spontaneous arousal and during the torpid state. Summer active bats were used as a control group. Rates of fractional protein breakdown were calculated from the rates of fractional protein synthesis and the net change in muscle and liver protein mass. Summer active bats were used as a control group. The mean duration of hibernation (torpor) states in this bat was 155 hours and the mean duration of arousals was 3.2 hours. During the hibernation season, the big brown bat spends 98% of overall time in torpor and 2% in arousals. Pre-hibernation the mean weights of the pectoralis muscle and the liver were 1.20 g and 0.73 g, respectively. After 110 days of hibernation,

Table 1 Protein metabolism in the big brown bat.

Tissue	Group	Fractional rate of protein synthesis (%/day)	Fractional rate of protein degradation (%/day)	Fractional rate of net protein loss (%/day)
Pectoralis muscle	Summer Active	8.8	≥ 9	≥ 0
	Aroused	2.2	22	20
	Hibernating	< 1	< 1	0.01
Liver	Summer Active	63.0	≥ 63	≥ 0
	Aroused	53.3	73	23
	Hibernating	< 1	< 1	0.01

the mean weights of the pectoralis muscle and the liver had fallen to 0.88 g (27% decrease), and 0.54 g (26% decrease), respectively. Table 1 summarizes protein metabolic parameters in the big brown bat. Yacoe does note that fractional rates of net protein loss and fractional rates of protein degradation during spontaneous arousals and in summer active bats are estimates only and that the potential for error is considerable. After 110 days of hibernation, mean total pectoralis muscle protein had fallen from 0.41 g to 0.22 g (46% decrease) and mean total liver protein from 0.24 g to 0.11 g (54% decrease). These decreases in pectoralis muscle and liver total protein account for 60% and 68% of the decreases in pectoralis muscle and liver mass, respectively. The 110 days of hibernation consisted of 107.8 days of torpor (98% of overall time) and 2.2 days of arousals (2% of overall time). Using the fractional rates of net protein loss given in Table 1, a 44% decrease in pectoralis muscle total protein and a 51% decrease in liver total protein would occur during arousals. Hence in this small mammal, during a 110 day hibernation season about 95% of the pectoralis muscle (44/46) and liver protein (51/54) losses occur during periods of arousal (about two days of total time) and only 5% during periods of torpor (about 108 days of total time). This data are consistent with arousal as a period of protein catabolism probably to supply amino acids as a substrate for gluconeogenesis. Since net losses of protein accounted for 60%–68% of the decreases in pectoralis muscle and liver mass, it seems reasonable to assume that the muscle and liver atrophy occurred predominantly during periods of arousal. Assuming the changes in protein metabolism described by Yacoe in the big brown bat are generalizable, then the following picture emerges for small mammal hibernators. During periods of hibernation (torpor), protein metabolism is virtually suspended and very little muscle atrophy occurs. During periods of arousal, significant protein turnover occurs with the rate of degradation exceeding the rate of synthesis. Most of the muscle atrophy observed in the hibernator would occur during these catabolic arousal states. However, protein balance would also be influenced by any food intake after arousal and before the next bout of hibernation. Although it has been claimed that muscle oxidative capacity (using citrate synthase activity as a measure) is increased during hibernation (Wickler *et al.*, 1987 and 1991; Steffen *et al.*, 1991), this conclusion is not supported if one uses total tissue citrate synthase activity rather than mass specific activity.

Natural animal model: the American black bear

The American black bear is a very appropriate natural animal model for disuse muscle atrophy in the human. As already discussed, small mammal hibernators appear to suffer little muscle atrophy during periods of torpor because their metabolic rate and rate of protein turnover are markedly suppressed. However, during short periods of arousal protein turnover significantly increases and since the rate of degradation exceeds that of synthesis loss of muscle mass occurs. The techniques of hindlimb immobilization and suspension for inducing muscle atrophy are unnatural and the results differ not only between these two techniques but also from the effects of bed rest in humans. In contrast, the bear remains inactive for five to seven months per year, has no intake during this time and experiences no spontaneous arousals. In addition, lean body mass does not significantly change during hibernation compared to pre-hibernation values (Lundberg *et al.*, 1976; Barboza *et al.*, 1997). Koebel *et al.* (1991) measured biochemical parameters in bear skeletal muscle (gastrocnemius and extensor hallucis longus) prior to denning (October), during the later stages of denning (March) and following spring arousal (May). Mass specific protein content and mass specific citrate synthase activity were not significantly different between the three observation time points. Since whole muscle weights were unknown the investigators could not document the actual extent of any muscle atrophy. However, based upon changes in mass specific DNA content and in the protein/DNA ratio between the three measurement times, the investigators estimated that the hibernating bear suffered about 10%–20% muscle atrophy. Since mass specific protein concentration was the same prior to and following hibernation, the muscle atrophy was not due to loss of water.

Tinker *et al.* (1998) measured both histological and biochemical parameters in bears during the late fall (early denning period) and in late winter/early spring (late denning period). The mean sampling interval for all bears was 123 days. A comparison of measurements made in the late denning versus early denning period gave the following results. Both the biceps femoris and gastrocnemius muscles showed a modest decrease in mass specific protein content amounting to about 10% and 4%, respectively. The percent water content of the biceps femoris increased significantly during hibernation (mean change 2.68%) while the water content in the gastrocnemius did not differ significantly. There was a significant decrease in the mean percentage of slow-twitch fibers in the biceps femoris (9.9%) but a non-significant change (3.56%) in the mean percentage of slow-twitch fibers in the gastrocnemius muscle. The CSAs of fast- and slow-twitch fibers in both muscles were not significantly altered between the early and late denning periods and as well there was no significant change in the mean number of total (combined fast- and slow-twitch) fibers counted per 3.9 mm^2 in either muscle. Mass specific citrate synthase activity decreased significantly in the biceps femoris but did not show a significant alteration in the gastrocnemius muscle.

In a recent study, Harlow *et al.* (2001) measured non-invasively the strength (peak torque force) of the tibialis anterior muscle in wild bears in late autumn, just after denning, and in early spring before they emerged from their dens. Mean muscle strength decreased by 23% between these two observation times. According to the study by Berg *et al.* (1997), humans after six weeks of bed rest lose 25%–30% muscle strength of the knee extensors tested in a comparable manner. Harlow *et al.* extrapolated this result in humans to predict a loss of strength of 90% over 130 days (the approximate period of hibernation in the bear). However such an extrapolation may not be valid since we know that in rodents subjected to HS or HI, the decreases in muscle weight and muscle fiber CSA and the change in relative proportions of type 1 and type 2 fibers are not linear with time (Booth, 1977; Thomason and Booth, 1990).

The bear: a perspective

Hindlimb suspension and hindlimb immobilization in animals (usually rodents are the test animal) and bed rest in humans induce significant muscle atrophy measured as a decrease in muscle weight and/or a reduction in muscle fiber size. In small mammal hibernators, almost all of the muscle atrophy occurs during periods of spontaneous arousal. Little atrophy occurs during periods of torpor since protein turnover (synthesis and degradation) is profoundly reduced. The hibernating bear appears to be an outlier among mammals. Although Koebel *et al.* (1991) estimated muscle atrophy of between 10%–20% in the denning bear; this value was based on very indirect measurements. The data of Tinker *et al.* (1998) included measurements of fiber size (CSA) and fiber number. Neither of these two parameters was altered by 123 days of hibernation (inactivity). These data are strong evidence that the hibernating bear experiences very little if any muscle atrophy. During hibernation, the bear maintains a rate of overall protein synthesis equivalent to or greater than the rate of protein synthesis during active months (Lundberg *et al.*, 1976; Barboza *et al.*, 1997) and has a rate of overall protein degradation about equal to that of synthesis (Barboza *et al.*, 1997). Since skeletal muscle constitutes approximately 53% of fat free mass (Tinker *et al.*, 1998), it is reasonable to assume that these measures of protein synthesis/degradation also reflect rates of muscle protein synthesis/degradation. The mechanisms/signals by which inactivity results in reduced muscle protein synthesis and increased muscle protein degradation are poorly understood. Somehow the bear has modified these mechanisms/signals. The 23% loss of muscle strength during hibernation does not appear to be due specifically to a loss of contractile proteins since there was no change in fiber size (types 1 and 2). The underlying basis then, for the muscle weakness is unclear. Berg *et al.* (1997) noted the same discrepancy in humans in whom six weeks of bed rest resulted in a reduction of muscle strength of 25%–30% whereas the magnitude of muscle atrophy was only 14%.

A number of research questions could be considered. However, the technical difficulties of performing experiments on such a large mammal as the bear are clearly considerable. Given this constraint, the following are some possibilities.

(1) Measurement of protein turnover (synthesis/degradation) in muscle prior to hibernation, at several time points during hibernation and post-hibernation. Liu and Barrett (2002) have recently reviewed methods for measuring protein metabolism *in vivo*. Such measurements would give direct experimental evidence to support (or possibly refute) the hypothesis that during hibernation in the bear, rates of muscle protein synthesis and degradation remain coupled and similar in magnitude to the rates found in active bears.

(2) Measurement of muscle size using non-invasive radiological techniques. I am unsure if portable ultrasound can be used to measure the size of specific muscles at least in terms of width and length. If so, then these measurements should be made several times during the period of hibernation as well as pre- and post-hibernation. If feasible, such measurements would give a direct assessment of the extent of any muscle atrophy occurring during hibernation and the time course if detectable atrophy does indeed occur.

(3) The maintenance of fiber size during hibernation implies that contractile proteins are not lost. Yet muscle strength decreases by 23%. The reason for this discrepancy has not been explored, although there are no suggestions for specific experiments to address this issue.

It seems therefore, that the key to the bear's successful hibernation is its ability to completely recycle urea, re-utilize the urea nitrogen for amino acid synthesis and maintain a rate of protein synthesis equal to or greater than the rate during active months. These metabolic characteristics enable the hibernating bear to maintain bone and muscle mass as well as avoid the complications of renal failure. The general features of protein metabolism in the hibernating bear are reviewed in Chapter 2, and in that chapter a number of research questions are discussed relating to this topic. As already pointed out, the adaptations of the hibernating bear should not be compartmentalized into isolated sets dealing with "renal failure," disuse osteoporosis and disuse muscle atrophy. All of these adaptations are part of an integrated whole. Another important question previously considered is whether the bear has evolved unique metabolic characteristics such as the capacity to synthesize essential amino acids (see Chapter 2) or evolved unique molecules which for example could be responsible for regulating bone homeostasis, or has the bear just modified and extended mechanisms and biomolecules present in all mammals. For want of any direct experimental evidence to the contrary, the latter possibility seems much more likely to me. The microflora in the bear's gastrointestinal tract could be the source of essential amino acids as has been demonstrated in other mammals (see Chapter 2). In addition, a known molecule such as parathyroid hormone may be the elusive osteoregulatory substance proposed by Floyd *et al.* (1990) to be present in the hibernating bear. Although there is much we do not know about the bear, the available data clearly indicate that biological solutions do exist for the problems of disuse osteoporosis and disuse muscle atrophy.

References

Al-Badry, K., Taha, H.M., 1983. Hibernation-hypothermia and metabolism in hedgehogs, changes in water and electrolytes. *Comp. Biochem. Physiol.* 74A, 435–441.

Barboza, P.S., Farley, S.D., Robbins, C.T., 1997. Whole-body urea cycling and protein turnover during hyperphagia and dormancy in growing bears (*Ursus americanus* and *U. arctos*). *Can. J. Zool.* 75, 2129–2136.

Berg, H.E., Larsonn, L., Tesch, P.A., 1997. Lower limb skeletal muscle function after 6 weeks of bed rest. *J. Appl. Physiol.* 82, 182–188.

Booth, F.W., 1977. Time course of muscular atrophy during immobilation of hind limbs in rats. *J. Appl. Physiol.* 43, 656–661.

Booth, F.W., 1982. Effect of limb immobilization on skeletal muscle. *J. Appl. Physiol.* 52, 1113–1118.

Brown, D.C., Mulhausen, R.O., Andrew, D.J., Seal, U.S., 1971. Renal function in anaesthetized dormant and active bears. *Am. J. Physiol.* 220, 293–298.

Clouston, W.M., Lloyd, H.M., 1987. Immobilation-induced hypercalcemia and regional osteoporosis. *Clin. Orthop. Rel. Res.* 216, 247–252.

David, M., Nelson, R., Alt, G., 1988. Study of calcium, phosphorous, alkaline phosphatase, and hydroxyproline in black bears. *FASEB J.* 2, A843 (abstract #3180).

Donahue, S.W., Vaughan, M.R., Demers, L.M., Donahue, H.J., 2003a. Serum markers of bone metabolism show bone loss in hibernating bears. *Clin. Orthop. Rel. Res.* 408, 295–301.

Donahue, S.W., Vaughan, M.R., Demers, L.M., Donahue, H.J., 2003b. Bone formation is not impaired by hibernation (disuse) in black bears. *J. Exp. Biol.* 206, 4233–4239.

Donahue, S.W., McGee, M.E., Harvey, K.B., Vaughan, M.R., Robbins, C.T., 2006a. Hibernating bears as a model for preventing disuse osteoporous. *J. Biomech.* 39, 1480–1488.

Donahue, S.W., Galley, S.A., Vaughan, M.R., Patterson-Buckendahl, P., Demers, L.M., Vance, J.L., McGee, M.E., 2006b. Parathyroid hormone may maintain bone formation in hibernating black bears (*Ursus americanus*) to prevent disuse osteoporosis. *J. Exp. Biol.* 209, 1630–1638.

Doty, S.B., Nunez, E.A., 1985. Activation of osteoclasts and the repopulation of bone surfaces following hibernation in the bat, *Myotis lucifugus*. *Anat. Rec.* 213, 481–495.

Feldman, D., Malloy, P.J., Gross, C., 2001. Vitamin D: biology, action, and clinical implications. In: Marcus, R., Feldman, D., Kelsey, J. (Eds.), *Osteoporosis*, 2nd Edn., Vol. 1. Academic Press, San Diego, California, pp. 257–303.

Fitts, R.H., Metzger, J.M., Riley, D.A., Unsworth, B.R., 1986. Models of disuse: a comparison of hindlimb suspension and immobilization. *J. Appl. Physiol.* 60, 1946–1953.

Fitzpatrick, L.A., Heath III, H., 2001. Primary hyperparathyroidism and hyperparathyroid bone disease. In: Marcus, R., Feldman, D., Kelsey, J. (Eds.), *Osteoporosis*, 2nd Edn., Vol. 2. Academic Press, San Diego, California, pp. 259–269.

Floyd, T., Nelson, R.A., Wynne, G.F., 1990. Calcium and bone metabolic homeostasis in active and denning black bears (*Ursus americanus*). *Clin. Orthop. Rel. Res.* 255, 301–309.

Friedman, P.A., 2000. Renal calcium metabolism. In: Seldin, D.W., Giebisch, G. (Eds.), *The Kidney, Physiology and Pathophysiology*, 3rd Edn., Lippincott Williams and Wilkins, Philadelphia, PA, pp. 1749–1789.

Gokhale, J.A., Robey, P.G., Boskey, A.L., 2001. The biochemistry of bone. In: Marcus, R., Feldman, D., Kelsey, J. (Eds.), *Osteoporosis*, 2nd Edn., Vol. 1. Academic Press, San Diego, California, pp. 107–188.

Goldspink, D.F., 1977. The influence of immobilization and stretch on protein turnover of rat skeletal muscle. *J. Physiol.* 264, 267–282.

Gordeladze, J.O., Drevon, C.A., Syversen, U., Reseland, J.E., 2002. Leptin stimulates human osteoblastic cell proliferation, *de novo* collagen synthesis, and mineralization: impact on differentiation markers, apoptosis, and osteoclastic signaling. *J. Cell. Biochem.* 85, 825–836.

Haller, A.C., Zimny, M.L., 1977. Effects of hibernation on interradicular alveolar bone. *J. Dent. Res.* 56, 1552–1557.

Harlow, H.J., Lohuis, T., Beck, T.D.I., Iaizzo, P.A., 2001. Muscle strength in overwintering bears. *Nature* 409, 997.

Harlow, H.J., Lohuis, T., Anderson-Sprecher, R.C., Beck, T.D.I., 2004. Body surface temperature of hibernating black bears may be related to periodic muscle activity. *J. Mammal.* 85, 414–419.

Harvey, K.B., Donahue, S.W., 2004. Bending properties, porosity, and ash fraction of black bear (*Ursus americanus*) cortical bone are not compromised with aging despite annual periods of disuse. *J. Biomech.* 37, 1513–1520.

Harvey, K.B., Drummer, T.D., Donahue, S.W., 2005. The tensile strength of black bear (*Ursus americanus*) cortical bone is not compromised with aging despite annual periods of hibernation. *J. Biomech.* 38, 2143–2150.

Hellgren, E.C., Vaughan, M.R., Kirkpatrick, R.L., Scanlon, P.F., 1990. Serial changes in metabolic correlates of hibernation in female black bears. *J. Mammal.* 71, 291–300.

Hellgren, E.C., Rogers, L.L., Seal, U.S., 1993. Serum chemistry and hematology of black bears: physiological indices of habitat quality or seasonal patterns? *J. Mammal.* 74, 304–315.

Hellgren, E.C., 1998. Physiology of hibernation in bears. *Ursus* 10, 467–477.

Hock, J.M., 2001. Anabolic actions of PTH in the skeletons of animals. *J. Musculoskel. Neuron Interact.* 2, 33–47.

Holloway, W.R., Collier, F.M., Aitken, C.J., Myers, D.E., Hodge, J.M., Malakellis, M., Gough, T.J., Collier, G.R., Nicholson, G.C., 2002. Leptin inhibits osteoclast generation. *J. Bone Miner. Res.* 17, 200–209.

Hudson, N.J., Franklin, C.E., 2002. Maintaining muscle mass during extended disuse: aestivating frogs as a model species. *J. Exp. Biol.* 205, 2297–2303.

Kiratli, J., 2001. Immobilization osteopenia. In: Marcus, R., Feldman, D., Kelsey, J. (Eds.), *Osteoporosis*, 2nd Edn., Vol. 2. Academic Press, San Diego, California, pp. 207–227.

Koebel, D.A., Miers, P.G., Nelson, R.A., Steffen, J.M., 1991. Biochemical changes in skeletal muscles of denning bears (*Ursus americanus*). *Comp. Biochem. Physiol.* 100B, 377–380.

Kwiecinski, G.G., Krook, L., Wimsatt, W., 1987. Annual skeletal changes in the little brown bat, *Myotis lucifugus*, with particular reference to pregnancy and lactation. *Am. J. Anat.* 178, 410–420.

Leblanc, A.D., Schneider, V.S., Evans, H.J., Engelbretson, D.A., Krebs, J.M., 1990. Bone mineral loss and recovery after 17 weeks of bed rest. *J. Bone Miner. Res.* 5, 843–850.

Lee, C.A., Einhorn, T.A., 2001. The bone organ system: form and function. In: Marcus, R., Feldman, D., Kelsey, J. (Eds.), *Osteoporosis*, 2nd Edn., Vol. 1. Academic Press, San Diego, California, pp. 3–20.

Lian, J.B., Stein, G.S., 2001. Osteoblast biology. In: Marcus, R., Feldman, D., Kelsey, J. (Eds.), *Osteoporosis*, 2nd Edn., Vol. 1. Academic Press, San Diego, California, pp. 21–71.

Liu, Z., Barrett, E.J., 2002. Human protein metabolism: its measurement and regulation. *Am. J. Physiol. Endocrinol. Metab.* 383, E1105–E1112.

Lundberg, D.A., Nelson, R.A., Wahner, H.W., Jones, J.D., 1976. Protein metabolism in the black bear before and during hibernation. *Mayo Clin. Proc.* 51, 716–722.

Martin, K.J., Gonzalez, E.A., Slatopolsky, E., 2004. Renal osteodystrophy. In: Brenner, B.M. (Ed.), *Brenner and Rector's The Kidney*. 7th Edn. Saunders, Philadelphia, PA, pp. 2255–2304.

Martin, T.J., Rodan, G.A., 2001. Coupling of bone resorption and formation during bone remodeling. In: Marcus, R., Feldman, D., Kelsey, J. (Eds.), *Osteoporosis*, 2nd Edn., Vol. 1. Academic Press, San Diego, California, pp. 361–371.

Mayer, W.V., Bernick, S., 1958. A comparative study of the dentin, periodontium aqnd alveolar bone of warm and awake and hibernating arctic ground squirrels, *Spermophilus undulatus*. *Anat. Rec.* 131, 580.

Milbury, P.E., Vaughan, M.R., Farley, S., Matula Jr, G.J., Convertino, V.A., Matson, W.R., 1998. A comparative bear model for immobility-induced osteopenia. *Ursus* 10, 507–520.

Nelson, R.A., Jones, J.D., Wahner, H.W., McGill, D.B., Code, C.F., 1975. Nitrogen metabolism in bears: urea metabolism in summer starvation and in winter sleep and role of urinary bladder in water and nitrogen conservation. *Mayo Clin. Proc.* 50, 141–146.

Nicks, D.K., Beneke, W.M., Key, R.M., Timson, B.F., 1989. Muscle fiber size and number following immobilization atrophy. *J. Anat.* 163, 1–5.

Nissenson, R.A., 2001. Parathyroid hormone and parathyroid hormone-related protein. In: Marcus, R., Feldman, D., Kelsey, J. (Eds.), *Osteoporosis*, 2nd Edn., Vol. 1. Academic Press, San Diego, California, pp. 221–246.

Overstreet, M., Floyd, T., Polotsky, A., Hungerford, D.S., Frondoza, C.G., 2003. Enhancement of osteoblast proliferative capacity by growth factor-like molecules in bear serum. *In Vitro Cell Dev. Biol.-Animal* 39, 4–7 (January and February 2003).

Overstreet, M., Floyd, T., Polotsky, A., Hungerford, D.S., Frondoza, C.G., 2004. Induction of osteoblast aggregation, detachment, and altered integrin expression by bear serum. *In Vitro Cell. Dev. Biol.-Animal* 40, 4–7 (January and February 2004).

Pardy, C.K., Wohl, G.R., Ukrainetz, P.J., Sawers, A., Boyd, S.K., 2004. Maintenance of bone mass and architecture in denning black bears (*Ursus americanus*). *J. Zool. Lond.* 263, 359–364.

Patel, M.S., Karsenty, G., 2001. Mouse genetics as a tool to study bone development and physiology. In: Marcus, R., Feldman, D., Kelsey, J. (Eds.), *Osteoporosis*, 2nd Edn., Vol. 1. Academic Press, San Diego, California, pp. 213–219.

Powers, S.K., Kavazis, A.N., DeRuisseau, K.C., 2005. Mechanisms of disuse muscle atrophy: role of oxidative stress. *Am. J. Physiol. Regul. Integr. Comp. Physiol.* 288, R337-R344.

Reid, M.B., 2005. Response of the ubiquitin-proteasome pathway to changes in muscle activity. *Am. J. Physiol. Regul. Integr. Comp. Physiol.* 288, R1423–R1431.

Rodriguez, M., Nemeth, E., Martin, D., 2005. The calcium-sensing receptor: a key factor in the pathogenesis of secondary hyperparathyroidism. *Am. J. Physiol. Renal Physiol.* 288, F253–F264.

Ross, F.P., Teitelbaum., 2001. Osteoclast biology. In: Marcus, R., Feldman, D., Kelsey, J. (Eds.), *Osteoporosis*, 2nd Edn., Vol. 1. Academic Press, San Diego, California, pp. 73–105.

Steffen, J.M., Koebel, D.A., Musacchia, X.J., Milsom, W.K., 1991. Morphometric and metabolic indices of disuse in muscles of hibernating ground squirrels. *Comp. Biochem. Physiol.* 99B, 815–819.

Takata, S., Yasui, N., 2001. Disuse osteoporosis. *J. Med. Invest.* 48, 147–156.

Templeton, G.H., Padalino, M., Manton, J., Glasberg, M., Silver, C.J., Silver, P., DeMartino, G., Leconey, T., Klug, G., Hagler, H., Sutko, J.L., 1984. Influence of suspension hypokinesia on rat soleus muscle. *J. Appl. Physiol.* 56, 278–286.

Templeton, G.H., Sweeney, H.L., Timson, B.F., Padalino, M., Dudenhoeffer, G.A., 1988. Changes in fiber composition of soleus muscle during rat hindlimb suspension. *J. Appl. Physiol.* 65, 1191–1195.

Thomason, D.B., Biggs, R.B., Booth, F.W., 1989. Protein metabolism and B-myosin heavy-chain mRNA in unweighted soleus muscle. *Am. J. Physiol. Regul. Integr. Comp. Physiol.* 257, R300–R305.

Thomason, D.B., Booth, F.W., 1990. Atrophy of the soleus muscle by hindlimb unweighting. *J. Appl. Physiol.* 68, 1–12.

Tinker, D.B., Harlow, H.J., Beck, T.D.I., 1998. Protein use and muscle-fiber changes in free-ranging, hibernating black bears. *Physiol. Zool.* 71, 414–424.

Tucker, K.R., Seider, M.J., Booth, F.W., 1981. Protein synthesis rates in atrophied gastrocnemius muscles after limb immobilization. *J. Appl. Physiol.* 51, 73–77.

Uebelhart, D., Demiaux-Domenech, B., Roth, M., Chantraine, A., 1995. Bone metabolism in spinal cord injured individuals and in others who have prolonged immobilization. A review. *Paraplegia* 33, 669–673.

Whalen, J.P., Krook, L., Nunez, E.A., 1972. A radiographic and histologic study of bone in the active and hibernating bat (*Myotis lucifugus*). *Anat. Rec.* 172, 97–108.

Wickler, S.J., Horwitz, B.A., Kott, K.S., 1987. Muscle function in hibernating hamsters: a natural analog to bed rest? *J. Therm. Biol.* 12, 163–166.

Wickler, S.J., Hoyt, D.F., Van Breukelen, F., 1991. Disuse atrophy in the hibernating golden-mantled ground squirrel, *Spermophilus lateralis*. *Am. J. Physiol.* 261, R1214–1217.

Yacoe, M.E., 1983. Protein metabolism in the pectoralis muscle and liver of hibernating bats, *Eptesicus fuscus*. *J. Comp. Physiol.* 152, 137–144.

Zerwekh, J.E., Ruml, L.A., Gottschalk, F., Pak, C.Y.C., 1998. The effects of twelve weeks of bed rest on bone histology, biochemical markers of bone turnover, and calcium homeostasis in eleven normal subjects. *J. Bone Miner. Res.* 13, 1594–1601.

CHAPTER 5

Ammonia Toxicity

Ammonia is a toxic molecule and more specifically it is a neurotoxin (Cooper and Plum, 1987). Mammals have low blood ammonia levels and develop neurological complications when these levels become elevated. There are a number of causes of hyperammonemia in humans (Mathias *et al.*, 2001) with the most common being acute and chronic liver failure. On the other hand, some species of fish are able to tolerate much higher blood ammonia levels than mammals without experiencing toxicity. First, the evidence for ammonia tolerance in fish and relevant features of overall ammonia metabolism in mammals and fish are reviewed. However, the major thrust of this chapter will be an examination of what is known about the mechanism of ammonia neurotoxicity in mammals and what is known about adaptations in the fish which enable them to tolerate ammonia concentrations that are toxic in mammals.

Ammonia Tolerance and Metabolism

Ammonia tolerance

Ammonia exists in two forms: a non-ionic species (NH_3) and an ionic species (NH_4^+). The pK of the NH_3/NH_4^+ system in blood is about 9.1–9.2 (Cooper and Plum, 1987) so that in biological fluids the ratio of NH_4^+ to

NH_3 will be about 100. In most studies, ammonia concentrations refer to total ammonia (NH_3 and NH_4^+) and in this chapter, unless otherwise stated, the term ammonia will denote the sum of NH_3 and NH_4^+. In general, most biological membranes are permeable to the non-ionic NH_3, but relatively impermeable to the charged NH_4^+ species (Randall and Tsui, 2002). Since ammonia is a neurotoxin, a consideration of how this substance crosses the blood-brain barrier (BBB) is important. Ammonia enters the brain from the blood by diffusion and not by a saturable transport system (Cooper and Plum, 1987; Felipo and Butterworth, 2002a). The uncharged species, NH_3 crosses the BBB more easily than the ionic form NH_4^+ but since the concentration of NH_4^+ is so much higher in the blood than that of NH_3, approximately 25% of ammonia enters the brain as NH_4^+. Assuming that intracerebral ammonia diffuses as readily as blood ammonia and mixes freely with it and given the evidence that ammonia crosses the BBB by diffusion (mainly as NH_3), then the concentration of ammonia in the brain should depend on the pH gradient between blood and brain (Cooper and Plum, 1987). Assuming a blood pH of 7.4 and a brain intracellular pH of 7.1, these considerations would lead to a prediction of a normal brain to blood concentration ratio of 2 (Cooper and Plum, 1987). It is worth emphasizing that this two fold higher brain ammonia concentration compared to blood is explicable by a modified Henderson-Hasselbalch equation based on the previously listed assumptions (Cooper and Plum, 1987). This prediction is close to published ratios which range from 1.5 to 3.0 (Cooper and Plum, 1987). In addition, the increased brain ammonia concentration observed in liver failure is probably not the result of a change in the integrity of the BBB with increased diffusion of ammonia from blood to brain, although the evidence for this conclusion is mixed. Induction of acute liver failure by portacaval shunting plus hepatic artery ligation in rats caused no increase in the permeability of the BBB (Mans *et al.*, 1994). However, positron emission tomography in cirrhotic patients with mild encephalopathy has given results consistent with an increase in the brain permeability/surface area product (Felipo and Butterworth, 2002a). In summary, under normal circumstances, brain ammonia concentration is a function of diffusion of ammonia from the blood plus formation of ammonia in the brain via several enzymatic pathways including the reversible oxidative deamination of glutamate by glutamate dehydogenase (Felipo and Butterworth, 2002a).

Table 1 contains a summary of ammonia concentrations in the blood and brain of mammals and fish. In general, normal blood ammonia levels

Table 1 Ammonia concentrations in blood (μM) and brain (μmol/g).

A	Normal values	Blood	Brain	Reference(s)
	Human	43.7–113 (arterial)		Cooper and Plum (1987) Clemmesen *et al.* (2000) Lockwood *et al.* (1979) Deferrari *et al.* (1981)
	Rat	50–250 (arterial)	0.222–0.794	Cooper and Plum (1987) Dejong *et al.* (1993) Chamuleau *et al.* (1994) Kosenko *et al.* (1994) Mans *et al.* (1994) Swain *et al.* (1992)
	Dog	19.4–119 (arterial)		Weber and Veach (1979) Fine (1982) Richards *et al.* (1971)
	Pig	25 (arterial) 75 (arterial) after a protein meal		Welters *et al.* (1999)
	Elasmobranch (*Raja erinacea*)	370		Ballantyne (1997)
	Swamp eel (*Monopterus albus*)	300	1.0	Ip *et al.* (2004b)
	Gulf toadfish (*Opsanus beta*)	250	1.5	Wang and Walsh (2000)
	Rainbow trout (*Oncorhynchus mykiss*)	400–653 1050 after feeding		Wicks and Randall (2002) Randall and Wright (1987)
	Mudskippers *Periophthalmodon schlosseri* *Boleophthalmus boddaerti*		1.5 2.0	Ip *et al.* (2005)
B	**Hyperammonemic states**	**Blood**	**Brain**	**Reference(s)**
	Human; cirrhosis with chronic encephalopathy.	149 (arterial)		Lockwood *et al.* (1979)
	Human; acute liver failure with severe encephalopathy. Time of sampling patients after development of liver failure unknown.	182 (arterial)		Clemmesen *et al.* (2000)

Table 1 (*Continued*)

B Hyperammonemic states	Blood	Brain	Reference(s)
Human; comatose neonate with ornithine transcarbamylase deficiency.	1420		Mathias *et al.* (2001)
Rat; acute liver failure (portacaval shunt with common bile duct ligation) with encephalopathy day 7 after induction.	188	1.1	Dejong *et al.* (1993)
Rat; acute liver ischemia (portacaval shunt plus celiac trunk ligation) Time of sampling rats unknown.		1.95	Chamuleau *et al.* (1994)
Rat; acute liver ischemia (portacaval shunt plus ligation of hepatic artery), all the rats had severe encephalopathy and death usually occurred within ten hours. Rats sampled within six hours of surgery.	1080	0.936	Mans *et al.* (1994)
Rat; toxicity induced by intraperitoneal injection of ammonium acetate. Rats sampled 15 minutes after injection. Half of the rats developed convulsions within the 15 minutes.	2700	4.00	Kosenko *et al.* (1994)
Rat; acute liver ischemia (portacaval shunt plus ligation of hepatic artery), rats developed severe encephalopathy and were sampled within eight to 12 hours of surgery		5.43	Swain *et al.* (1992)
Mouse; a specifically generated mouse model which is completely deficient in carbamoyl phosphate synthetase, the first enzyme of the urea cycle. These mice die within 12–36 hours of birth.	4565		Schofield *et al.* (1999)
Rainbow trout; (*Oncorhynchus mykiss*) exposed to environmental ammonia level causing 50% mortality in 96 hours.	1882–2176		Wicks and Randall (2002)

Table 1 (*Continued*)

B	Hyperammonemic states	Blood	Brain	Reference(s)
	Gulf toadfish; (*Opsanus beta*) exposed to a sub-lethal environmental ammonia level	1250	4.5	Wang and Walsh (2000)
	Mudskippers			Ip *et al.* (2005)
	Periophthalmodon schlosseri		18.0	
	Boleophthalmus boddaerti		14.5	
	Both exposed to a sub-lethal environmental ammonia level			
	Swamp eel (*Monopterus albus*) exposed to a sub-lethal environmental ammonia level	3500	6.48	Ip *et al.* (2004b)

in mammals are 100 μM or less although Cooper and Plum (1987) report a concentration of 250 μM in the rat. Normal brain ammonia concentrations are between 0.2 and 0.8 μmol/g. Fish, on the other hand, have baseline blood levels considerably higher and in the rainbow trout blood levels can exceed 1000 μM after feeding. Baseline brain concentrations in the fish range between 1 and 2 μmol/g. In mammals, encephalopathy can occur with ammonia blood levels in the range of 200 μM and brain concentrations of 1–2 μmol/g. In the rat with acute liver failure, death occurs with plasma levels of about 1000 μM and brain concentrations of about 0.9–5.43 μmol/g (Table 1, Mans *et al.*, 1994; Swain *et al.*, 1992). It is interesting that Mans *et al.* (1994) observed that in rats with acute liver failure, plasma ammonia levels correlated better with cerebral dysfunction than brain ammonia levels. Death also occurs in the mouse with complete carbamoyl phosphate synthetase deficiency and plasma ammonia levels in the range of 5000 μM. In contrast, fish appear to tolerate high ammonia concentrations at least on an acute basis. For example, the swamp eel can survive with blood levels as high as 3500 μM and in mudskippers, brain concentrations as high as 18 μmol/g are not lethal. These concentrations are in the lethal range for mammals (Table 1). In fact, the brain ammonia concentrations in the mudskippers are about 3–4.5 times higher than the levels measured by Kosenko *et al.* (1994) in rats made ammonia toxic by intraperitoneal injection of ammonium acetate or in rats with acute liver ischemia (Swain *et al.*, 1992). In the studies involving gulf toadfish, fish exposed to ammonia were much less active and less responsive to stimuli in comparison with the control groups.

The greater ammonia tolerance of fish compared to other vertebrates is supported by the toxicity experiments of Wilson *et al.* (1968 and 1969). These investigators measured the intraperitoneal LD50 values for ammonium acetate for mouse, chicken and three fish species (rainbow trout, channel catfish and goldfish). The mean values in mmol/kg body weight were: mouse 10.84, chicken 10.44, rainbow trout 17.74, channel catfish 25.73, and goldfish 29.34.

Since the central nervous system is the site of ammonia toxicity and since fish tolerate much higher brain ammonia levels than mammals, then the adaptations underlying this tolerance must be intrinsic to the central nervous system.

Ammonia metabolism

Ammonia is generated in both mammals and fish primarily from deamination of the alpha amino nitrogen of amino acids principally through the coupled action of transaminases and glutamate dehydrogenase (Campbell, 1991; Ballantyne, 2001). Another source of ammonia is through deamination of adenosine monophosphate (AMP) by the purine nucleotide cycle which in the mammal, is found in muscle, brain and kidney tissues (Campbell, 1991) and in fish is found at least in muscle (Ballantyne, 2001). In fish it has been estimated that the liver is responsible for between 50%–70% of ammonia production (Ballantyne, 2001). However, hepatectomized eels continue to have daily ammonia excretion rates unchanged from controls suggesting that amino acid deamination can occur in extrahepatic sites (Singer, 2003). Ammonia blood levels will reflect the balance between production on the one hand and excretion/detoxification on the other. Blood ammonia levels rise significantly after feeding as noted in Table 1 for the pig and rainbow trout. Welters *et al.* (1999) demonstrated that in the pig, the source of the increased blood ammonia after a protein meal was increased renal ammoniagenesis with a greater renal vein ammonia efflux. The increased renal ammonia production did not appear to be a response to changes in acid-base homeostasis. In the rainbow trout, the postprandial increase in blood ammonia reflects increased ammonia production from liver and probably also muscle (Wicks and Randall, 2002).

In mammals, interorgan ammonia metabolism has been reviewed by Huizenga *et al.* (1996) and Olde Damink *et al.* (2002). Interorgan ammonia metabolism is a complex topic and what follows is a very simplified discussion. As already noted, ammonia is generated mainly from deamination of the alpha amino nitrogen of amino acids. Mammalian tissues initially detoxify ammonia by converting it to glutamine which can then serve as a vehicle for transferring nitrogen to other tissues including the liver. However, several organs in the mammal do add ammonia directly into the blood. The intestinal tract is a major exporter of ammonia to the liver via the portal venous circulation. The bulk of this ammonia is a metabolic product of the action of glutaminase on the amide nitrogen of glutamine. Within the liver there is compartmentalization of ammonia handling (Olde Damink *et al.*, 2002; Young and Ajami, 2001). In the periportal hepatocytes, ammonia delivered by the portal circulation and ammonia released within the hepatocyte by the action of glutaminase on glutamine or the action of glutamate dehydrogenase on glutamate, is fed into the ornithine-urea cycle. The perivenous hepatocytes are rich in glutamine synthetase and hence any ammonia escaping detoxification through the ornithine-urea cycle can be metabolized to glutamine. During exercise muscle produces ammonia as a result of increased activity of the purine nucleotide cycle. Although glutamine synthetase activity is low in skeletal muscle, by virtue of its mass, muscle is one of the principal glutamine synthesizing organs. The kidney also releases ammonia into the systemic circulation and glutamine is the main substrate for renal ammoniagenesis in rats, dogs and humans. In the pig alanine appears to be the likely substate for renal ammonia production (Welters *et al.*, 1999). Brain ammonia metabolism will be considered in a following section, but in general the brain demonstrates net uptake of ammonia (Cooper and Plum, 1987; Olde Damink *et al.*, 2002). In mammals then, intestine, kidney and (during exercise) muscle are the principal ammonia producing organs. Glutamine and to some extent alanine serve as the main vehicles for interorgan nitrogen transfer (Young and Ajami, 2001; Rodwell, 2000) while in the fish, as will be described in the next section, ammonia itself functions as an important carrier molecule for nitrogen (Singer, 2003). Hence, systemic blood levels of ammonia are low in the mammal because ammonia is transported in the blood chiefly as glutamine and alanine and ammonia delivered to the liver either directly or indirectly as glutamine and alanine is detoxified through the synthesis of urea.

As already noted, in fish the liver is the principal ammonia producing organ. Production of ammonia results from amino acid catabolism with the primary mechanism being transdeamination as in the mammal (Ballantyne, 2001). The purine nucleotide cycle contributes little to ammonia production in the liver (Ballantyne, 2001). Muscle is also a source of ammonia and it has been estimated that the release of ammonia from glutamate by the action of glutamate dehydrogenase accounts for about two thirds of the ammonia production of goldfish muscle with the purine nucleotide cycle accounting for the remainder (van Waarde, 1981). There is evidence in the fish that renal ammoniagenesis with renal excretion of ammonia plays a role in acid-base homeostasis (Dantzler, 1989). However it is not known whether the kidney adds ammonia to the blood as it does in the mammal. Plasma glutamine levels are lower in fish than in mammals (Ballantyne, 2001). Glutamine synthetase levels are much lower in fish muscle than in mammalian muscle (Ballantyne, 2001). In fact, studies on the species and tissue distribution of glutamine synthetase in fish have generally shown high levels in the brain and little or no activity in the liver of most fish except species of teleosts that are ureogenic and ureo-osmotic elasmobranchs (Anderson, 2001). In fish muscle glutamine does not serve as a nitrogen store for ammonia under normal conditions as it does in mammals. Synthesis of glutamine in tissues such as muscle for export to other tissues (as occurs in the mammal) does not occur in fish (Ballantyne, 2001). In fish muscle, for example, glutamine is used as a fuel source (Ballantyne, 2001). In fish, ammonia replaces glutamine as an important vehicle for intertissue nitrogen transport. Most teleosts (bony fish) are ammonotelic, that is ammonia generated in the liver and other tissues is not detoxified by "fixation" (for example, through synthesis of glutamine) but is simply excreted directly across the gills into the surrounding water medium (Anderson, 2001). Gill ammonia excretion is predominantly a combination of passive NH_3 and NH_4^+ diffusion (Singer, 2003). In contrast, marine elasmobranchs and some teleosts (for example the gulf toadfish, *Opsanus beta*) have an active ornithine-urea cycle and excrete urea rather than ammonia (Anderson, 2001). However, as summarized in Table 1, the gulf toadfish is still capable of tolerating blood and ammonia concentrations much higher than the mammal. Hence, despite the ability of this fish to detoxify ammonia through the synthesis of urea, its central nervous system still possesses an intrinsic tolerance to high ammonia concentrations.

In summary, one of the major differences between mammals and fish is that under normal circumstances fish do not rely on tissue "fixation" of ammonia (for example, through synthesis of glutamine) for detoxification of this molecule. The major exception to this generalization is the brain in fish which as already noted contains high levels of glutamine synthetase. In most fish, direct excretion of ammonia into the surrounding water is the principal "detoxification" mechanism. Although fish have evolved various strategies to avoid high blood and brain ammonia concentrations under adverse environmental conditions (Ip *et al.*, 2004a), fish have also evolved the capacity to tolerate high blood and brain levels of ammonia. The data in Table 1 indicate that these adaptations reside within the central nervous system itself.

Brain Ammonia Metabolism

The following review of brain ammonia metabolism is based primarily on studies in mammals. However, the relatively limited data available for fish suggest that brain ammonia metabolism in this vertebrate approximates that of the mammal (Ballantyne, 2001; Soengas and Aldegunde, 2002).

As discussed in the "Ammonia Tolerance" section, the predicted brain/blood ammonia concentration ratio is 2, given that ammonia crosses the blood brain barrier by diffusion and that the pH of blood and brain cells are assumed to be 7.4 and 7.1, respectively. Published ratios, according to Cooper and Plum (1987), are in the range of 1.5 to 3; close to the predicted value. Cooper and Plum also point out that in conditions of sustained hyperammonemia brain ammonia concentrations are also about 1.5 to 3 times those of the blood, i.e. in a hyperammonemic state, given sufficient time, an equilibrium situation will be re-established and the brain to blood ammonia concentration will again be about 2 in accordance with the modified Henderson-Hasselbalch equation referred to in the "Ammonia Tolerance" section. In Table 1, many of the brain to blood concentration ratios, both in normal and hyperammonemic states are also in this range. Arterial-venous (AV) differences for ammonia have been measured across the brain of humans and other mammals (Cooper and Plum, 1987; Deferrari *et al.*, 1981; Porro and Maiolo, 1970; Webster and Gabuzda, 1958). The results have been inconsistent but as summarized in the paper by Cooper and Plum (1987) most of the studies give a positive value indicating net uptake of ammonia by the brain. Cooper and Plum also comment that in hyperammonemic animals, a positive AV difference for ammonia appears to be always present. Cooper and Plum (1987) also point out that ammonia diffuses from cerebrospinal fluid (CSF) into brain.

In the brain there are numerous enzymatic pathways that result in the production of ammonia. However, the most important are mediated by glutaminase, glutamate dehydrogenase and the purine nucleotide cycle (Cooper and Plum, 1987). Glutaminase is particularly abundant in the nerve endings of glutamatergic neurons and catalyzes the breakdown of glutamine to glutamate and ammonia. Glutamate dehydrogenase (present in both neurons and astrocytes) catalyzes the reversible oxidative deamination of glutamate and under normal or even moderately hyperammonemic conditions this reaction goes in the direction of ammonia production especially in the astrocyte (Cooper and Plum, 1987). The purine nucleotide cycle is also found in brain tissue (primarily astrocytes) and according to Cooper and Plum (1987) may be responsible for generating a major fraction of whole brain ammonia production.

Hence, ammonia diffuses into the brain from both the blood and CSF. In addition, ammonia is produced by enzymatic reactions within the brain, chiefly the ones discussed in the previous paragraph. Since excess ammonia is toxic to the central nervous system, its concentration in the brain is kept low by the activity of glutamine synthetase. This enzyme, localized in astrocytes, detoxifies incoming or endogenously generated ammonia by catalyzing the synthesis of glutamine from glutamate and ammonia (Cooper and Plum, 1987). In hyperammonemic states the AV difference for ammonia across the brain is consistently positive and in some reports much more positive than under normal physiological conditions (Webster and Gabuzda, 1958; Porro and Maiolo, 1970; Deferrari *et al.*, 1981; Dejong *et al.*, 1993). In liver disease the hyperammonemia is the result of reduced ornithine-urea cycle enzyme activity due to hepatocyte loss as well as to the occurrence of portal-systemic shunting of blood such that ammonia exported by the intestine bypasses the liver and spills into the systemic circulation. In children with inborn errors of ornithine-urea cycle enzymes, the hyperammonemia is due to a reduction in the urea synthesizing capacity of the liver. In these conditions systemic blood ammonia levels will be high as will brain ammonia concentrations. The elevated blood ammonia levels will clearly lead to the diffusion of more ammonia into the brain. However what is not clear is whether these hyperammonemic states are also associated with increased brain ammonia production. For example, another cause of hyperammonemia is the hyperinsulinism/hyperammonia (HI/HA) syndrome. This syndrome is caused by mutations of glutamate dehydrogenase that reduce the sensitivity of this enzyme to allosteric inhibition by the high energy phosphates, GTP and ATP (Stanley, 2004). Children with this syndrome have blood ammonia concentrations between two and five times the normal concentration. The hyperammonia is due to excessive oxidative deamination of glutamate in the liver. HI/HA patients do not suffer from encephalopathy but do have an increased frequency of seizure activity (Stanley, 2004). One might predict that these patients would have increased brain ammonia levels due to increased diffusion of ammonia from blood to brain plus increased brain ammonia production as a result of excessive oxidation of glutamate by glutamate dehydrogenase. If an animal model of this syndrome were available, the effect of a loss of inhibitory control of glutamate dehydrogenase on brain ammonia production could be measured.

In summary, under normal physiological conditions, brain ammonia concentration is a function of addition of ammonia through diffusive movement from blood and CSF, production of ammonia within brain tissue and removal of ammonia via synthesis of glutamine from glutamate and ammonia. In humans, the most common cause of hyperammonemia is acute/chronic liver disease. In this condition, increased systemic blood ammonia levels occur as a result of decreased hepatic detoxification of intestinal derived ammonia through the synthesis of urea. In fish, hyperammonemia usually results from environmental conditions such as a high ambient water pH or high ambient water ammonia concentrations (Randall and Tsui, 2002).The latter can occur due to agricultural run-off and decomposition of biological waste. In these conditions ammonia excretion by the gills is impaired. In addition, when ambient water ammonia levels are very high, there can be net uptake of ammonia by the fish from the surrounding water. In hyperammonemic states, blood and brain concentrations of ammonia are increased. Clearly high blood ammonia levels will result in increased diffusive ammonia uptake by the brain, but what is not apparent is whether increased brain ammonia production also occurs in hyperammonemic states. Finally, brain ammonia metabolism is an integral component of the so-called "glutamate-glutamine" cycle which is central to the synthesis of the neurotransmitter glutamate.

Current explanations for the molecular mechanism of ammonia toxicity are largely based upon the interactions between the neurotransmitter glutamate and its receptors and how these interactions are affected by ammonia. In the next section, the properties of glutamatergic neurons and the "glutamate-glutamine" cycle will be reviewed.

Glutamatergic Neurons

General considerations (Purves *et al.*, 2004)

The synapses within the brain can be generally divided into two different classes: electrical synapses and chemical synapses. At electrical synapses, ions flow through specialized channels which bridge the pre- and postsynaptic plasma membranes. In contrast, chemical synapses facilitate cell to cell communication via the secretion of neurotransmitters. Transmission at a chemical synapse involves an elaborate sequence of events. Small-molecule neurotransmitters, such as glutamate, are synthesized within the cytoplasm of the presynaptic terminals and then loaded into synaptic vesicles via transporters in the vesicular membrane. When an action potential arrives at the presynaptic terminal, depolarization of the presynaptic terminal membrane causes opening of voltage-gated Ca^{2+} channels. There is a rapid influx of calcium into the terminal which triggers fusion of the synaptic vesicles with the plasma membrane of the presynaptic neuron. neurotransmitters are released into the synaptic cleft via exocytosis. The neurotransmitters molecules diffuse across the synaptic cleft and bind to specific receptors on the postsynaptic neuron membrane. There are two broad families of receptors: ionotropic and metabotropic. Ionotropic receptors contain two functional domains: an extracellular site that binds neurotransmitters and a membrane spanning domain that forms an ion channel. Metabotropic receptors do not have ion channels as part of their structure; instead they affect the properties of channels by the activation of intermediate molecules called G proteins. Neurotransmitter binding activates G proteins which then dissociate from the receptor and interact either directly with ion channels or indirectly via other intracellular messengers. With both types of receptors, neurotransmitter binding triggers the associated channels to open or sometimes to close. The subsequent change in ion flows through these channels results in a change in the postsynaptic membrane potential (PSP). If the change in the PSP increases the likelihood of a postsynaptic action potential occurring then the PSP is called excitatory. If the change in the PSP decreases the likelihood of a postsynaptic action potential occurring then the PSP is called inhibitory. In addition, it should be noted that the binding of neurotransmitters to specific receptors not only leads to an electrical response of the postsynaptic cell, but in many cases activates additional intracellular pathways within the postsynaptic cell that have a variety of functional consequences.

Glutamate, a Neurotransmitter and the "Glutamate-Glutamine" Cycle

Nearly all excitatory neurons in the central nervous system are glutamatergic and it is estimated that over half of all brain synapses release this neurotransmitter (Purves *et al.*, 2004). Ammonia, glutamate and glutamine metabolism are compartmented within the brain (Cooper, 2001). Glutamate is synthesized within neurons with the chief precursor being glutamine, which is released by astrocytes. Glutamine is taken up into the presynaptic nerve terminal and metabolized to glutamate by the mitochondrial enzyme glutaminase (Purves *et al.*, 2004). The glutamate synthesized within the cytoplasm is then packaged into synaptic vesicles. Once released into the synaptic cleft, glutamate is taken up by the perineuronal astrocytes via high affinity transporters. Astrocytes express the glutamate transporters, GLT-1 and GLAST (Butterworth, 2002). GLT-1 is widely distributed throughout the brain whereas GLAST is expressed mainly in the cerebellum (Butterworth, 2002). Within the astrocyte, glutamate is converted into glutamine by the enzyme glutamine synthetase and the glutamine is then retransported out of the astrocyte and into the presynaptic nerve terminal. This movement of glutamine from astrocyte to neuron and glutamate in the reverse direction is commonly referred to as the "glutamine-glutamate" cycle. One turn of this cycle results in the movement of a molecule of ammonia from the astrocyte to the nerve terminal (Butterworth, 2001).

Two major classes of glutamate receptors have been identified on the postsynaptic neuron membrane: ionotropic and metabotropic (Felipo and Butterworth, 2002a).

Metabotropic receptors do not directly gate ion channels. Binding of glutamate to these receptors activates G proteins which modulate ion channels as well as recruit second messengers such as phospholipase C and D and adenyl cyclase (Felipo and Butterworth, 2002a).

The three ionotropic receptors are called NMDA (N-methyl-D-aspartate) receptors, AMPA (α-amino-3-hydroxy-5-methyl-isoxazole-4-propionate) receptors and kainate (kainic acid) receptors — named after the agonists that activate them (Purves *et al.*, 2004). Activation of these ionotropic receptors always produces excitatory postsynaptic responses. Members of each receptor group differ in kinetics, voltage dependencies and regulation and the properties of the synaptic response therefore depend on which specific receptor types are assemblied at the postsynaptic site. Of these three, the NMDA receptor system has been the one most thoroughly studied with respect to the effects of ammonia (Felipo and Butterworth, 2002a). NMDA receptors have a Mg^{2+} binding site in the pore of the channel. Under normal conditions cations cannot enter the postsynaptic neuron because Mg^{2+} is blocking the NMDA receptor channel. When the postsynaptic neuron depolarizes, Mg^{2+} is pushed out of the pore, allowing other ions to flow through the channel. Release of this magnesium block by synaptic depolarization is a necessary first step. Activation of the NMDA receptor depends on the simultaneous binding of two agonists, glutamate released from synaptic vesicles and glycine from the extrasynaptic fluids. The activation of NMDA receptors results in increased flow of Na^+, K^+ and Ca^{2+} through the ion channel (Monfort et al., 2002). The increased flow of calcium leads to increased intracellular calcium which binds to calmodulin and activates different enzymes. One of the enzymes activated is neuronal nitric oxide synthase thus leading to increased formation of nitric oxide. Nitric oxide activates soluble guanylate cyclase leading to increased synthesis of cyclic guanosine monophosphate (cGMP) (Monfort et al., 2002). In addition, the binding of calcium to calmodulin activates the protein phosphatase calcineurin which dephosphorylates Na^+/K^+ ATPase thus increasing its activity.

Glutamate receptors are not limited to neurons but are also found on glial cells (Teichberg, 1991; Backus et al., 1989). Pharmacological techniques have been used in cultures of astrocytes from rat cerebral hemispheres to characterize these receptors. AMPA and kainite receptors have been detected, but astrocytes do not express NMDA receptors. In addition, the experimental results suggested that the kainate and AMPA receptors were not independent and that in the cultured astrocyte, kainate and AMPA activated a common receptor complex (Backus et al., 1989). In neurons, these two receptor types are independent. There is also evidence that astrocytes express metabotropic glutamate receptors (Teichberg, 1991).

The properties of glutamatergic synapses described in the preceding paragraphs are based primarily on studies done in mammals. However, glutamatergic synapses have also been studied in non-mammals such as the lamprey (Soengas and Aldegunde, 2002; Shupliakov et al., 1996; Buchanan et al., 1987; Shupliakov et al., 1997). In the lamprey, the giant reticulospinal synapse represents a central glutamatergic synapse with properties similar to those of mammalian brain synapses. The "glutamate-glutamine" cycle has been demonstrated in this model. Shupliakov et al. (1997) using quantitative immunocytochemical techniques, showed the presence of glutamate predominantly in synaptic vesicles within the axons and glutamine concentrated in small glial cell bodies and in glial processes which appeared to surround neuronal cell bodies and axons. The investigators also detected glutamine in neuronal cell bodies but the labeling decreased dramatically within the axons especially the more distal portions away from the cell bodies. In another study, Shupliakov et al. (1996) reported prominent uptake of glutamate into glial processes encapsulating each synaptic junction. In addition these investigators discuss the role of several proteins, synapsin and synaptotagmin, in the process of vesicle fusion and release of neurotransmitter contents in the lamprey model system. Similar proteins are involved in the process of vesicular release of neurotransmitters in the mammal. Buchanan *et al.* (1987) demonstrated the existence of NMDA and non-NMDA postsynaptic glutamatergic receptors in the lamprey. Using the lamprey reticulospinal glutamatergic system, these investigators applied selective antagonists of these receptors and observed a depression of the late component of the excitatory postsynaptic potential.

The molecular biology of the NMDA receptor has been studied in mammals and in several species of fish (Purves *et al.*, 2004; Dunn *et al.*, 1999; Cox *et al.*, 2005; Harvey-Girard and Dunn, 2003). In the mammal, the NMDA receptor is a tetrameric protein complex containing two different types of subunit, NR1 and NR2. The functional receptor is assembled from two NR1 and two NR2 subunits. There are multiple subtypes of both NR1 and NR2. In mammals, there are five NMDA receptor genes: one encoding subunit NR1 (with different subtypes generated through the utilization of alternative RNA splicing) and four encoding four NR2 subunits (NR2A/B/C/D). Different combinations of NR1 and NR2 can form functional receptor complexes but their properties will vary depending on the subunits in the complex. The NR1 subunit has a wide distribution throughout the central nervous system, whereas the NR2 subunits exhibit distinct regional and cell specific expression patterns. The NR1 subunit carries the binding site for the co-agonist glycine and the asparagine residue involved in the voltage-dependent magnesium block is located in the second transmembrane domain on all subunits. Dunn *et al.* (1999) have studied the NR1 subunit in the teleost *Apteronotus leptorhynchus*. The amino acid sequence of aptNR1 is about 88% identical to the amino acid sequence of NR1 subunits from frog, duck, rat and human. The most highly conserved segment includes the proposed transmembrane segments, the pore segment and the ligand binding domains. In the zebrafish, Cox *et al.* (2005), reported two NR1 genes and eight NR2 genes. The two zebrafish NR1 subunits are approximately 90% identical to the human protein. The highest degree of homology is shown by the region containing the putative transmembrane sections as well as the magnesium binding domain. Whereas the zebrafish NR2 subunits do not exhibit as high an overall identity with the human subunits as do the two NR1s, all the subunits were approximately 80% identical through the region containing the putative transmembrane domains. Harvey-Girard and Dunn (2003) have studied the functional properties of NMDA receptors reconstituted from the teleost *Apteronotus leptorhynchus*. To study the teleost NMDA receptor, the AptNR1 and AptNR2B cDNAs were co-expressed in the HEK cell line for electro-physiological and pharmacological studies. To compare the properties of the teleost receptor with the mammalian receptor, similar measurements were made with mouse NR1/NR2B receptors expressed in the HEK cell line. As already discussed, AptNR1 has an amino acid sequence about 88% identical to that of mammals. AptNR2B has an overall value of 62% amino acid identity compared to the mouse with the transmembrane segments and the ligand binding regions sharing the highest levels of sequence identity. Application of the agonist glutamate or NMDA in the presence of glycine produced robust currents with both Apt and m(mouse) receptors. Deactivation time courses were measured after NMDA pulses and the deactivation kinetics were similar for both the fish and murine receptors. The two receptors had similar relative calcium permeabilities but the fish NMDA receptor had a reduced affinity for Mg^{2+} compared to the murine NMDA receptor. Therefore the fish receptor would have a less complete Mg^{2+} block compared to the mammalian receptor. In terms of pharmacological measurements, both receptors showed similar affinities for the agonists glutamate and NMDA and the co-agonist glycine. In terms of antagonists, APV a competitive inhibitor for the glutamate site on NR2, had similar inhibitory effects in the fish and mouse NMDA receptor. However, the non-competitive antagonist ifenprodil had a significantly greater inhibitory effect in the mouse receptor compared to the fish receptor.

Hence, on the basis of these studies it is clear that the fish central nervous system has glutamatergic excitatory neurons and receptors analogous to those found in the mammal. The molecular biology of the NMDA receptor has been highly conserved across vertebrates. However, the study of Harvey-Girard and Dunn (2003) does identify several functional differences between the fish and mammalian NMDA receptor. These known differences and other, as yet unmeasured, differences may help to explain the higher tolerance of the fish brain to excess ammonia compared to the mammalian brain.

Mechanism of Ammonia Neurotoxicity

In humans, neurotoxicity due to excess ammonia is an important problem in diseases such as acute and chronic liver failure and in disorders of the enzymatic machinery responsible for the synthesis of urea. Fish can also suffer from excess ammonia but this state is generally the result of adverse environmental conditions which impair ammonia excretion. Based upon the data summarized in Table 1, it is clear that the central nervous system of fish is much more resistant to the toxic effects of ammonia than the central nervous system of mammals. A number of experimental studies have been done in mammals trying to elucidate the molecular mechanisms involved in ammonia neurotoxicity. Although excess ammonia has been shown to lead to a number of adverse effects, a consensus appears to be emerging that an important target of ammonia is the glutamatergic excitatory neuron. In this section, the effects of excess ammonia on the mammalian central nervous system are reviewed. To simplify the discussion, the effects are categorized into three groups: neuropathological effects, effects on brain energy metabolism and effects on glutamatergic synapses. Then what is known about the mechanisms underlying the tolerance of the fish central nervous system to levels of ammonia toxic to mammals will be considered.

Neuropathological effects

Acute and chronic liver diseases in adults are associated with distinct neuropathological changes (Felipo and Butterworth, 2002a).

In acute liver failure, brain edema occurs chiefly due to swelling of astrocytes and their processes. Cell swelling may be severe enough to cause raised intracranial pressure and brain herniation. Clemmesen *et al.* (1999) reported the characteristics of 44 patients with acute liver failure of whom 14 died of cerebral herniation. Arterial ammonia levels, measured within 24 hours after the development of grade III encephalopathy (pre-coma stage) were considerably higher in patients who died from cerebral herniation than in patients who did not (230 versus 118 μM, mean values). Similarly rats with acute ischemic liver failure (portacaval anastomosis and hepatic artery ligation) demonstrated a significant increase in brain water compared to sham operated control rats (Swain *et al.*, 1992). In addition, brain water content increased progressively with advancing stages of encephalopathy in the rats with acute ischemic liver failure. In the control rats, brain ammonia concentration was 0.750 μmol/g. In the rats with acute liver failure, brain ammonia levels rose from a mean value of 2.5 μmol/g in those with the least severe encephalopathy to a mean value of 5.43 μmol/g in those with the most severe encephalopathy. All measurements of brain water content and ammonia concentrations were made within eight to 12 hours of surgery. Rats made hyperammonemic by ammonium acetate also develop brain edema (Takahashi *et al.*, 1991). In this study, plasma ammonia levels went from a baseline of 31 μM (mean value) to 601 μM (mean value) after the intravenous infusion of ammonium acetate. Fractional brain water content increased significantly in the hyperammonemic rats by about 2.6% compared to control rats. This study demonstrates that acute hyperammonemia alone can lead to increased brain water content and hence the brain edema observed in acute liver disease can be attributed to the effects of hyperammonemia without incriminating other "toxins" which accumulate in liver failure.

In chronic liver failure, the morphological changes in astrocytes differ from those observed in acute liver failure (Felipo and Butterworth, 2002a; Cooper and Plum, 1987). The astrocytes are swollen although not to the same degree as in acute liver failure, but in addition show characteristic cytological alterations such as a large pale nucleus, prominent nucleolus and margination of their chromatin pattern. Astrocytes with this collection of cytological features are known as Alzheimer type II astrocytes.

Although changes in the astrocytes are the most frequently occurring neuropathological alterations observed in liver disease, neuronal cell loss has also been described (Felipo and Butterworth, 2002a; Cooper and Plum, 1987).

The mechanisms underlying ammonia induced neuropatholgical changes have been best studied for the increase in brain water content. Evidence has been gathered that the swelling of astrocytes (the cytological marker of the increase in brain water) in hyperammonemic states is the result of increased synthesis and accumulation of glutamine within astrocytes (Albrecht and Dolinska, 2001; Cooper, 2001). In rats with acute ischemic liver failure, brain glutamine levels increase more than five fold compared to control animals (Chamuleau *et al.*, 1994; Swain *et al.*, 1992). For example in the study by Swain *et al.* (1992), the mean brain glutamine concentration in rats with acute ischemic liver failure was 24 μmol/g compared to a mean value of 4.4 μmol/g in control sham operated rats. In the study by Takahashi *et al.* (1991) previously referred to in this section, cortical glutamine concentrations increased from 5.6 μmol/g (mean value) in control rats to 18.8 μmol/g (mean value) in rats which received the intravenous infusion of ammonium acetate. In this same study, Takahashi *et al.* (1991) treated a group of rats with L-methionine sulfoximine (MSO), a glutamine synthetase inhibitor, prior to the intravenous infusion of ammonium acetate. In these rats, the mean plasma ammonium concentration reached a value of 908 μM after the ammonium acetate infusion, but the prior administration of MSO prevented completely both the increase in cortical glutamine concentration and the increase in fractional brain water content that occurred when ammonium acetate was infused without prior treatment with MSO. Takahashi *et al.* (1991) suggested that in hyperammonemic states, there is increased synthesis of glutamine within astrocytes (glutamine synthetase is limited almost exclusively to astrocytes; Felipo and Butterworth, 2002a) and that the accumulated glutamine represented an increase in intracellular osmoles. Water flow into the astrocytes would increase and the astrocytes would swell. Brusilow and Maestri (1996) have proposed that in hyperammonemic states, astrocyte swelling as a result of intracellular glutamine accumulation not only leads to cerebral edema but also to astrocyte dysfunction. The latter would manifest as alterations in neurotransmitter and energy metabolism and ion transport just to name a few. According to these investigators astrocytic swelling provides an integrated explanation for the clinical, physiologic, neurochemical and pathologic consequences found in acute hyperammonemic encephalopathy, whether in patients with urea cycle disorders or patients with fulminant hepatic failure. A similar hypothesis has been advanced by Albrecht and Dolinska (2001) who quote evidence that glutamine accumulation in hyperammonemic states is responsible not only for cerebral edema but also for impairments in cerebral energy metabolism and alterations in cerebral blood flow.

Although the experimental evidence supporting a role for glutamine in the genesis of astrocytic swelling in hyperammonemic states seems strong, there are experimental observations consistent with other explanations. Vogels *et al.* (1997) studied the effect of memantine, a noncompetitive antagonist of the NMDA receptor, in rats with hyperammonemic encephalopathy. Two different experimental groups were studied: rats with a portacaval shunt subsequently given an intravenous ammonium acetate infusion (AI-PCS rats) and rats with acute ischemic liver failure (portacaval shunt followed by ligation of the hepatic artery; LIS rats). Memantine was given to the AI-PCS rats one hour after the start of the ammonium acetate infusion and was given to the LIS rats after loss of the righting reflex. In control experiments, saline (the vehicle for the memantine) was administered instead of memantine. In both groups of rats, arterial blood ammonia concentrations increased significantly reaching values of $1000 \, \mu$M in the AI-PCS group and $2000 \, \mu$M in the LIS group. Memantine did not affect blood ammonia levels in either group. The AI-PCS rats developed encephalopathy during the six hours of the ammonium acetate infusion while the LIS rats developed encephalopathy within three to six hours of ligation of the hepatic artery. The severity of the encephalopathy was quantified by clinical grading and EEG characteristics. In both groups, memantine administration significantly improved the severity of the encephalopathy. In both experimental groups, intracranial pressure (ICP) and brain water content increased. Memantine administration decreased ICP and brain water content in the AI-PCS rats, but had no significant effect on ICP and brain water content in LIS rats. In the AI-PCS group, memantine did not reduce brain water content completely back to control levels. These experimental results implicate the glutamatergic neuron/synapse, and more specifically the NMDA receptor in the pathogenesis of hyperammonemic encephalopathy, and to some extent in the pathogenesis of cerebral edema. However, the mechanism by which memantine reduces brain water may be mediated by effects on non-NMDA receptors as has been demonstrated for the NMDA receptor channel blocker, MK-801. (See discussion of effects of extracellular glutamate on astrocyte swelling in the paragraph following the next one.) Why memantine was effective in reducing the increase in brain water content in the AI-PCS group but not the LIS group is not clear. The investigators do discuss this particular point and advance several explanations.

Several other experimental studies implicate yet other factors in the pathogenesis of the cerebral edema in hyperammonemic states. Margulies *et al.* (1999) studied the role of the aquaporin-4 (AQP4) water channel in rats with fulminant hepatic failure (FHF). Rats with surgically induced FHF developed a significant increase in brain cortex water content 36 hours postinduction of FHF compared to control rats. Increased mRNA and protein levels of AQP4 were measured in the cortex of the 36-hour postFHF group compared to the control group. In a somewhat complementary study, Desjardins *et al.* (2001) assessed the role of the glucose transporter Glut 1 in the genesis of cerebral edema in a rat model. Glut 1 is a member of the family of glucose transporters and is found in the brain on perivascular endothelial cells and astrocytes. Glut 1 functions as a glucose transporter and water channel protein in brain. Acute ischemic liver failure was induced by portacaval anastomosis followed by hepatic artery ligation. Six hours after the induction of liver failure and prior to the development of cerebral edema, Glut 1 mRNA levels were increased by 71% and this was accompanied by increased Glut 1 immunolabeling of endothelial cells and pericapillary astrocytes throughout the cerebral cortex and basal ganglia compared to control rats. The onset of brain edema was accompanied by a further increase in Glut 1 expression. In contrast, the expression of the neuronal glucose transporter Glut 3 was unchanged at all time points during the progression of acute liver failure.

In hyperammonemic states, brain extracellular glutamate concentrations are increased probably due to a decrease in expression of astrocyte glutamate transporters resulting in reduced uptake of glutamate by astrocytes (Felipo and Butterworth, 2002a). Exposure of cultured rat cortical astrocytes to high concentrations of glutamate results in morphological changes including an increase in intracellular water (Bridges *et al.*, 1992; Chan *et al.*, 1990). Glutamate at a concentration of 1 mM caused a transient increase in intracellular water space. The swelling peaked at eight hours and then returned to normal by 24 hours. Higher concentrations of glutamate (10 mM) caused progressive swelling of astrocytes over 24 hours with evidence of actual cell damage. Quisqualate an agonist of the AMPA receptor, also induced swelling of cultured astrocytes but NMDA and kainite receptor agonists did not. The increase in astrocyte intracellular water is most likely due to an increase in ion (Na^+) influx through activated (opened) glutamate receptor ion channels followed by osmotic driven flow of water. Interestingly, MK-801, a non-competitive channel blocker of the NMDA receptor was effective in completely preventing the astrocyte swelling induced by glutamate and quisqualate (Chan *et al.*, 1990). Since these cultured astrocytes lack the NMDA receptor, MK-801 must be blocking ion channels that are independent of NMDA receptors.

In summary, the genesis of astrocyte swelling (cerebral edema) in hyperammonemic states involves the interplay of multiple factors/processes. Intracellular accumulation of glutamine is one of those factors/processes, but as the sole mechanism it is probably not sufficient to account for the astrocyte swelling or for that matter the behavioral/neurological characteristics of hyperammonemic encephalopathy. The other factors/processes involved in astrocyte swelling include; decreased expression of astrocyte glutamate receptors and increased expression of water channel proteins. The signals linking excess brain ammonia to increased expression of water channel proteins are presently unknown.

Effects on Brain Energy Metabolism

In their review article, Cooper and Plum (1987) summarized the effects of ammonia on brain energy metabolism. Brain ammonia levels in the range of 1–3 μmol/g are associated with falls in adenosine triphosphate (ATP) levels. However a mechanism underlying this effect is not presented. Ammonia stimulates glycolysis in brain extracts by activation of the enzyme phosphofructokinase. In addition, high concentrations of ammonia reduce brain glucose utilization (Cooper and Plum, 1987). There are several reports that ammonia interferes with the tricarboxylic acid cycle and Cooper and Plum cite evidence that ammonia inhibits the enzymes alpha-ketoglutarate dehydrogenase and isocitrate dehydrogenase. During glycolysis, which occurs in the cytoplasm, glucose is converted to pyruvate. This net oxidation results from the conversion of cytoplasmic NAD^+ to NADH at the glyceraldehyde-3-phosphate dehydrogenase step of glycolysis. For glycolysis to proceed, NAD^+ must be regenerated and this process is accomplished through oxidation of NADH by the mitochondrial respiratory chain. However cytoplasmic NADH cannot cross the mitochondrial membrane, so this process is accomplished by transfer of reducing equivalents across the mitochondrial membrane via substrate pairs linked by suitable dehydrogenases. The malate-aspartate shuttle is the vehicle by which these reducing equivalents (in this case malate as the source of 2H) are transported from the cytosol across the mitochondrial membrane. Since hyperammonemia is associated with decreases in brain glutamate and aspartate levels, it has been proposed that there will be a decreased flux through the malate–aspartate shuttle and thereby interference with normal oxidative metabolism via the tricarboxylic acid cycle and respiratory chain (Cooper and Plum, 1987). An extensive metabolic analysis was done by Mans et al. (1994) in rats with acute ischemic liver failure (portacaval anastomosis and ligation of the hepatic artery). These rats developed significant encephalopathy and death usually occurred between six and ten hours after surgery. Mean plasma ammonia and brain ammonia levels increased from 80 μM and 0.222 μmol/g in sham operated controls to 1080 μM and 0.936 μmol/g in rats with grade 4 encephalopathy. Some of the other changes measured were as follows (rats with acute liver failure versus sham operated controls): brain glutamine increased by 3.4 times, plasma lactate and brain lactate increased by 5.6 and 5.2 times respectively, whole blood glucose utilization was decreased by 15%–39%, brain glutamate concentration decreased by 29%, brain levels of ATP, adenosine diphosphate (ADP), adenosine monophosphate (AMP) and phosphocreatine were unchanged, and brain pyruvate levels increased by 2.6 times. The increase in lactate and pyruvate brain levels probably represents an increased rate of glycolysis relative to tricarboxylic acid cycle activity. Interestingly, Mans et al., (1994) found that plasma ammonia levels and brain glutamine levels correlated the best with the grade of encephalopathy and brain glucose utilization. Reduced brain concentration of glutamate was also reported in the studies of Takahashi et al. (1991) and Swain et al. (1992) previously discussed in the "Neuropathological Effects" section. Although whole brain glutamate concentration is decreased, there is also a redistribution of glutamate between intra-and extracellular compartments. Extracellular glutamate is increased due to a decrease in glutamate uptake by astrocytes as well as an increase in glutamate release from astrocytes (Chan and Butterworth, 2003; Felipo and Butterworth, 2002a) whereas intracellular glutamate concentration would be decreased. (See "Effects on Glutamatergic Synapses" next section.)

Some of the observations concerning changes in brain energy metabolites in hyperammonemic states may depend upon the severity of the ammonia insult. Kosenko *et al.* (1994) studied brain energy metabolism in rats rendered hyperammonemic by an intraperitoneal injection of ammonium acetate. Measurements were made 15 minutes after the ammonium acetate injection. Mean blood and brain ammonia concentrations in these rats were 2700 μM and 4.0 μmol/g compared to levels of 170 μM (blood) and 0.33 μmol/g (brain) in control rats. Hyperammonemia led to changes in other brain metabolites. The concentration of glucose, lactate, pyruvate and glutamine all increased, whereas the concentrations of glycogen, glutamate and ATP decreased. In this same study, Kosenko *et al.* (1994) also measured the effects of prior administration of MK-801, a blocker of the NMDA ion channel. MK-801 completely prevented the ammonia induced decrease in ATP levels and partially prevented the increases in glucose, lactate and pyruvate concentrations. On the other hand, prior administration of MK-801 amplified the ammonia induced changes in brain glutamine and glutamate levels, i.e. glutamine levels were higher and glutamate levels were lower after MK-801 and ammonium acetate injection than after ammonium acetate injection alone. Kosenko *et al.* (1994) demonstrated that this amplification of the effects of ammonia on brain levels of glutamine and glutamate by the prior injection of MK-801 is a reflection of the capacity of MK-801 to increase glutamine synthetase activity on its own. That MK-801, an NMDA receptor antagonist, completely prevents the ammonia induced depletion of ATP suggests that activation of this receptor mediates the increased consumption of ATP. Kosenko *et al.* (1994) did further experiments to explore the link between receptor activation and ATP depletion and these experiments will be discussed in the "Glutamatergic Synapses" section.

Effects on Glutamatergic Synapses

There is a growing body of evidence that the glutamatergic synapse plays a central role in mediating the neurotoxic effects of excess ammonia (Felipo and Butterworth, 2002a; Monfort *et al.*, 2002). Acute ammonia toxicity appears to be the result of activation of NMDA receptors by excess ammonia since NMDA receptor antagonists are very effective in preventing this toxicity. Hermenegildo *et al.* (1996) tested the protective effect in mice of 11 different NMDA receptor antagonists given 15 minutes before a lethal intraperitoneal injection of ammonium acetate. The antagonists consisted of channel blockers, competitive antagonists and compounds acting at the glycine binding site. Control mice given ammonium acetate only had a 95% mortality rate. All of the antagonists decreased the mortality rate although some were more effective than others. For example, MK-801, a non-competitive channel blocker, reduced the mortality rate to 25%, while mice given butanol, which acts at the glycine binding site, had a mortality rate of zero. These investigators also tested the protective effect of antagonists to non-NMDA receptors. An antagonist of AMPA receptors gave partial protection whereas an antagonist of kainite/AMPA receptors gave no protection. Antagonists to other neurotransmitter receptors (muscarinic and nicotinic acetylcholine receptors, GABA receptors) were either ineffective in preventing mortality or partially protective. In addition, mice given a prior injection of methionine sulfoximine, an inhibitor of the enzyme glutamine synthetase had a mortality rate of 100% following the injection of ammonium acetate. The results of this study are strong evidence that acute ammonia neurotoxicity is mediated primarily by activation of NMDA receptors by ammonia. The results of the study by Vogels *et al.* (1997), previously reviewed in the "Neuropathological Effects" section, also strongly support the hypothesis that ammonia induced activation of NMDA receptors underlies the genesis of encephalopathy in acute hyperammonemic states. These investigators demonstrated that memantine, an NMDA receptor channel blocker, significantly reduced the severity of hyperammonemia induced encephalopathy in two experimental rat models: rats with acute ischemic liver failure and rats with a portacaval anastomosis given an infusion of ammonium acetate.

The studies discussed in the previous paragraph indicate that the neurotoxic manifestations of acute hyperammonemic states can be prevented or ameliorated by antagonists of the NMDA receptor. The presumption of these studies is that the excess ammonia is responsible for the activation of NMDA receptors. Complementing these experimental results is a study by Hermenegildo *et al.* (2000) in which these investigators demonstrated activation of NMDA receptors by excess ammonia in rat brain in vivo. As reviewed in the "Glutamate, a Neurotransmitter and the

'Glutamate-Glutamine' Cycle" section, activation of NMDA receptors leads to increased intracellular Ca^{2+} in the postsynaptic neuron. The Ca^{2+} binds to calmodulin and activates nitric oxide synthase resulting in increased formation of nitric oxide. Nitric oxide activates guanylate cyclase leading to increased formation of cyclic guanosine monophosphate (cGMP). Rats were fitted with a microdialysis guide cannula in the cerebellum which allowed the investigators to sample brain (cerebellar) extracellular fluid. Rats were made hyperammonemic by an intraperitoneal injection of ammonium acetate and then samples of brain extracellular fluid were obtained for measurement of cGMP, ammonia and glutamate. The injection of ammonium acetate led to an increase in the extracellular concentration of cGMP in a dose dependent manner. In these rats, a dose of intraperitoneal ammonium acetate greater than 4 mmol/kg body weight was necessary to raise the extracellular concentration of cGMP, i.e. to activate NMDA receptors. The mean peak brain extracellular ammonia concentrations resulting in coma or death were 450 and 600 μM, respectively. There was a good correlation between the extracellular concentration of cGMP and the neurological effects (precoma, coma and death) induced by the ammonia injection. When rats were given MK-801, an NMDA receptor channel blocker, prior to the injection of ammonium acetate the rise in extracellular cGMP levels was completely prevented. To assess whether the ammonia induced activation of NMDA receptors was mediated by increased extracellular glutamate, the investigators determined the time course of changes in extracellular concentrations of ammonia, cGMP and glutamate following the injection of ammonium acetate. The maximum increase in the extracellular concentration of these metabolites occurred in the time sequence; ammonia first followed by cGMP second and then glutamate last. These observations indicate that ammonia is responsible for the activation of NMDA receptors and that the increased extracellular concentration of glutamate is primarily a consequence and not the cause of this activation. The ammonia induced activation of NMDA receptors is due to ammonia induced depolarization of the postsynaptic neuron (Binstock and Lecar, 1969) with release of the Mg^{2+} block allowing increased activation of the receptors without an increase in extracellular glutamate concentration.

The experimental observations described in the previous two paragraphs are very supportive of the construct that ammonia induced activation of NMDA receptors is the primary mechanism underlying acute ammonia neurotoxicity. The study by Hermenegildo *et al.* (2000) showed that ammonia itself was responsible for activation of the receptor but that a large dose of intraperitoneal ammonium acetate was required. Ammonia induced activation of NMDA receptors results in a number of associated events. As shown by Kosenko et al. (1994), one of these associated events is depletion of ATP. Activation of the receptor channel allows entry of Na^+ and Ca^{2+} ions into the neuron. To maintain sodium homeostasis the excess Na^+ entering through the channel is extruded from the cell by the action of Na^+, K^+-ATPase. Increased ATPase activity is a consequence of decreased phosphorylation brought about by an NMDA receptor mediated decrease of protein kinase C activity. In addition, the excess Ca^{2+} entering the cell through the opened receptor channels are taken up by mitochondria (Felipo and Butterworth, 2002b). The resulting impairment of mitochondrial function leads to decreased ATP synthesis. The removal of neuronally released glutamate from the synaptic cleft is achieved by high affinity high capacity energy dependent astrocytic glutamate transporters. Exposing cultured astrocytes to millimolar concentrations of ammonia results in a significant decrease in glutamate transporter mRNA and protein levels as well as a calcium dependent release of glutamate (Chan and Butterworth, 2003). These findings would certainly explain the elevated brain extracellular levels of glutamate observed in hyperammonemic states (Felipo and Butterworth, 2002a). Finally, the observations of Hermenegildo et al. (1996) and Kosenko et al. (1994) cast doubt on the importance of increased cellular levels of glutamine as a major factor in the manifestations of acute ammonia neurotoxicity. Hermenegildo et al. (1996) showed that prior administration of MK-801, an NMDA receptor channel blocker, reduced the mortality rate of a lethal intraperitoneal injection of ammonium acetate in mice from 95% to 25%. Kosenko et al. (1994) showed in rats given both MK-801 and an intraperitoneal injection of ammonium acetate that mean brain glutamine concentration was higher than in rats given MK-801 or ammonium acetate alone. In addition as already noted in the first paragraph of this section, Hermenegildo et al. (1996) showed that prior administration of methionine sulfoximine, an inhibitor of glutamine synthetase, was ineffective in reducing the mortality of a lethal intraperitoneal injection of ammonium acetate in their mice model.

The effects of ammonia on the glutamatergic synapse reviewed in the previous paragraphs refer to acute hyperammonemic states. Chronic hyperammonemia has different effects on this synapse and the differential effects of acute and chronic hyperammonemia on NMDA receptors have been summarized by Monfort et al. (2000). Since the ammonia toxicity studies performed in fish are generally of an acute nature, the effects of chronic hyperammonemia on the mammalian glutamatergic synapse will not be reviewed in this chapter.

Ammonia and the Fish Central Nervous System

There are many published studies on the topic of ammonia toxicity in fish. However there are only a limited number of studies which have addressed the question as to what properties of the fish central nervous system are responsible for ammonia tolerance. In fact, the differences between how fish and mammalian nervous systems interact with ammonia are just beginning to be characterized. Hence, currently there are insufficient data to paint a coherent picture describing the adaptations of the fish central nervous system which underlie its tolerance to high ammonia levels.

In general, hyperammonemia develops in fish when adverse environmental conditions occur which impede ammonia excretion via the gills. These adverse environmental conditions include a high ambient water pH or a high ambient water ammonia concentration. Ammonia is a neurotoxin in fish as it is in mammals. Ammonia excess in fish causes hyperventilation, hyperexcitability, coma, convulsions and finally death (Ip *et al.*, 2001). Fish have evolved a number of strategies to reduce or avoid ammonia toxicity and a description of these strategies has been the subject matter of several recent review articles (Ip *et al.*, 2001; Randall and Tsui, 2002). However, it is clear from the data in Table 1 that in addition to these strategies, the fish brain tolerates ammonia concentrations which are lethal in mammals. These strategies are listed rather than discussed in detail since the focus of this section is the adaptations (or at least what is known about them) that allow the fish brain to tolerate higher ammonia concentrations than mammals. The aforementioned strategies include: a reduction in ammonia production by decreasing the rates of proteolysis and amino acid catabolism, partial amino acid catabolism leading to the formation and storage of alanine, detoxification and storage of ammonia as glutamine and/or urea and active excretion of ammonia via the gills. Generally different fish species use different strategies although as noted by Tsui *et al.* (2004), the Oriental weatherloach uses many of these strategies. The detoxification of ammonia through the synthesis of glutamine is one strategy that requires further discussion. Glutamine is synthesized from glutamate and ammonia through the catalytic action of glutamine synthetase. As noted in the "Ammonia Metabolism" section, glutamine synthetase levels are much lower in fish muscle than in mammalian muscle and are also low in fish liver with the exception of those teleosts which are ureogenic and in ureo-osmotic elasmobranchs. High levels of glutamine synthetase are found in fish brain so detoxification of ammonia through synthesis of glutamine represents an important strategy for protecting the fish central nervous system from the toxic effects of ammonia. This topic will be discussed in more detail in this section.

Role of glutamine

In acute hyperammonemia, brain glutamine levels increase in both mammals and fish. As discussed in the "Neuropathological Effects" section, the synthesis and accumulation of glutamine in the mammalian brain, although important as a mechanism for detoxifying ammonia, has been implicated in the genesis of brain edema. Brain edema with cerebral herniation is a lethal complication of acute ammonia toxicity in mammals. However, in fish the story with respect to the effects of high brain glutamine levels appears to be different from that of mammals. Fish are resistant to high ammonia exposure despite the increase in brain glutamine. The ammonia tolerant swamp eel (*Monopterus albus*) when exposed to a high but sub-lethal concentration of environmental ammonia experiences a marked increase in both plasma and brain concentrations of ammonia as a result of impeded ammonia excretion (Table 1). This teleost possesses very high brain glutamine synthetase activity and during the period of ammonia exposure, the concentration of glutamine in the brain increased six fold (Ip *et al.*, 2004b). The central nervous system of this fish clearly has a high tolerance to ammonia at the cellular or subcellular level even in the face of high brain glutamine levels. However, the underlying mechanism responsible for this tolerance has not been defined. The detoxification of brain ammonia through the synthesis of glutamine is a subsidiary process helping to maintain brain ammonia levels within the tolerable range. The importance of this auxiliary process is supported by studies using fish of the family Batrachoididae (Wang and Walsh, 2000). These fish have a high ammonia tolerance predominantly due to an intrinsic, but as yet unknown, resistance of their nervous system to ammonia toxicity. The three members of the family Batrachoididae studied demonstrated a positive correlation between brain glutamine synthetase activity and the ability to tolerate ammonia as measured by a standard 96-hour LC50 test (lethal ambient water ammonia concentration to 50% of the population). This positive correlation underscores the importance of glutamine synthesis as a defense mechanism for ammonia toxicity in fish. A more definitive study was done by Veauvy *et al.* (2005). These investigators examined the effects of (ammonia induced) elevated brain glutamine levels on brain water content in the gulf toadfish (*Opsanus beta*). The gulf toadfish was one of the members of the family Batrachoididae studied by Wang and Walsh (2000). Magnetic resonance imaging was used to assess changes in brain water content in fish exposed to a sub-lethal environmental ammonia concentration. Brain glutamine levels were manipulated by using the glutamine synthetase inhibitor methionine sulfoximine (MSO). Ammonia exposure resulted in a three fold increase in brain ammonia concentration and a two fold increase in brain glutamine concentration. Pretreatment of the fish with MSO completely prevented the ammonia induced increase in brain glutamine concentration. There were no changes in brain water content in fish exposed to either sub-lethal or supra-lethal environmental concentrations of ammonia with or without pretreatment of the fish with MSO. Hence in fish unlike mammals, acute ammonia toxicity is not associated with brain edema. MSO is a toxic agent and fish treated with MSO alone showed 100% survival after 40 hours, 25% survival after 48 hours and 0% survival after 72 hours. The investigators expected that fish pretreated with MSO and exposed to a sub-lethal environmental concentration of ammonia would survive at least 40 hours. However, fish exposed to a sub-lethal level of ammonia and pretreated with MSO did not survive beyond 20 hours suggesting that glutamine synthetase activity is essential to delay the toxic effects of ammonia in the gulf toadfish. Ip *et al.* (2005) have also examined whether accumulation of glutamine in the brain mediates the toxicity of excess ammonia. These investigators used two species of mudskippers to study the effects of a prior injection of MSO on the mortality rate in fish injected with a lethal dose of ammonium acetate. The dose of MSO used decreased glutamine synthetase activity by 30% to 36%. In neither species did pretreatment with MSO reduce the mortality rate of a lethal dose of ammonium acetate. This observation is consistent with the construct that in fish, ammonia toxicity is not mediated by an increase in the concentration of brain glutamine.

In summary, mammals and fish differ in the apparent effects of ammonia induced accumulation of glutamine in the brain. In mammals the increased concentration of glutamine in the brain (or more specifically within the astrocytes) is believed to be an important factor in the genesis of brain (astrocyte) swelling. The accumulated glutamine acts as an osmolyte promoting increased bulk flow of water into the astrocytes. As discussed in the "Neuropathological Effects" section, several investigators have proposed that glutamine induced astrocyte swelling is the underlying mechanism responsible for the manifestations of ammonia toxicity. In contrast, the increased accumulation of glutamine in the brain of fish does not play a role in mediating the toxic effects of excess ammonia. In fish, the synthesis and accumulation of glutamine in the brain represents an important defense mechanism for the detoxification of excess brain ammonia.

Glutamate

Acute ammonia toxicity in rats is associated with an increase in brain glutamine concentration of about two to three fold and a decrease in brain glutamate concentration of between 28% and 44% (Mans et al., 1994; Swain et al., 1992; Takahashi et al., 1991; Kosenko et al., 1994). Decreases in brain glutamate concentration have also been measured in the mouse, cat and dog under conditions of hyperammonemia (Cooper and Plum, 1987). Presumably this decrease reflects the formation of glutamine from glutamate and ammonia through the action of glutamine synthetase. Acute ammonia toxicity studies in fish do not give such a consistent result. In the gulf toadfish, acute ammonia exposure resulted in a doubling of brain glutamine concentration and a decrease in brain glutamate concentration by about 25% (Veauvy et al., 2005). Glutamine synthetase activity was responsible for these changes in brain glutamine and glutamate concentration since these changes were prevented by pretreating the fish with MSO, an inhibitor of glutamine synthetase. In the swamp eel, acute ammonia exposure led to a six fold increase in brain glutamine concentration but no significant change in brain glutamate level (Ip et al., 2004b). In two species of mudskippers, brain glutamine concentration increased by four to 12 times during acute ammonia exposure but brain glutamate concentration did not significantly change (Ip et al., 2005). The mechanism by which these fish increase brain glutamine without depleting brain glutamate stores is unclear. In addition, the underlying biological significance of the differential response of fish and mammals with regard to ammonia induced changes in brain glutamate concentration is unknown. However, this difference is just additional evidence that the ammonia tolerance of fish resides in the biophysical properties of its central nervous system.

Brain energy metabolism

Veauvy *et al.* (2002) published an interesting study in which they compared the response of brain slices and isolated mitochondria from rats and gulf toadfish to millimolar concentrations of ammonia. In rat brain slices, ammonia induced a significant decrease (12%) in intramitochondrial NADH levels compared to control values, whereas ammonia exposure did not change intramitochondrial NADH levels in gulf toadfish brain slices. As discussed in the "Effects on Brain Energy Metabolism" section, ammonia inhibits several enzymes in the tricarboxylic acid cycle and reduces flux through the malate-aspartate shuttle. These effects would explain a reduction in intramitochondrial NADH levels in rat brain slices. The lack of an effect of ammonia on NADH levels in gulf toadfish brain slices suggests that ammonia does not affect brain energy metabolism in fish as it does in mammals. These investigators also examined the rates of oxidative phosphorylation for the electron chain enzymes in isolated brain mitochondria from rats and gulf toadfish. Ammonia had no effect on the enzymes of the electron chain in isolated mitochondria from either rats or gulf toadfish.

This study underscores the different responses of the fish and mammalian nervous systems to ammonia exposure. Since ammonia does not reduce intramitochondrial NADH levels in the fish brain then one would have to hypothesize that ammonia does not inhibit enzymes of the tricarboxylic acid cycle in fish as it does in mammals. Furthermore, since ammonia does not appear to cause a fall in brain glutamate concentration in fish, one could hypothesize that flux through the malate-aspartate shuttle is not decreased in fish as it is in mammals.

NMDA receptors

In mammals there is now strong evidence that activation of postsynaptic neuronal NMDA receptors by ammonia is the chief disturbance underlying the toxicity of ammonia. In contrast, there are very limited data in fish addressing the role of the glutamatergic synapse in mediating ammonia toxicity. Ip *et al.* (2005) examined the protective effect of MK-801, a non-competitive NMDA channel blocker in mudskippers injected with a lethal dose of ammonium acetate. The two species of mudskippers (*Periophthalmodon schlosseri* and *Boleophthalmus boddaerti*) used in these experiments are very ammonia tolerant (see Table 1). The prior administration of MK-801 did not have any protective effect in *P. schlosseri* or *B. boddaerti* given a lethal dose of ammonium acetate. In fact in one of the species, B. bod-daerti, mortality increased from 70% in controls to 100% in fish pretreated with MK-801. However, the same batch of MK-801 at the same dosage decreased the mortality (from 80% to 50%) in goldfish (Carassius auratus) injected with a lethal dose of ammonium acetate. Hence in these mud-skippers, activation of NMDA receptors by ammonia does not appear to be the underlying mechanism in acute ammonia toxicity. Tsui et al. (2004) examined the protective effect of MK-801 in the ammonia tolerant Oriental weatherloach (Misgurnus anguillicaudatus). Administration of MK-801 prior to the injection of a lethal dose of ammonium acetate reduced the mortality to 0% compared to 60% when fish were not given MK-801. Hence in this fish, it appears that activation of NMDA receptors is involved in ammonia toxicity.

It is difficult to reconcile the studies of Ip et al. (2005) and Tsui et al. (2004) other than to conclude that activation of NMDA receptors by ammonia is species dependent. Could it be that activation of NMDA receptors is a more important mechanism of toxicity in fish that are less ammonia tolerant? The ammonia tolerance of P. schlosseri, B. boddaerti and M. anguillicaudatus can be compared by using the measured 96-hour LC50s which are (in total ammonia concentrations): 120 mM, 13.5 mM and 75 mM, respectively (Ip et al., 2005; Tsui et al., 2004). MK-801 has a protective effect in M. anguilli-caudatus but not in P. schlosseri or B. boddaerti. Hence, there is no apparent correlation between degree of ammonia tolerance and activation of NMDA receptors. One more qualification needs to be considered. As discussed in the "Neuropathological Effects" section, MK-801 may not be completely specific for NMDA receptors but may also block ion channels associated with non-NMDA glutamate receptors.

As discussed in the "Effects on Glutamatergic Synapses" section, ammonia activates NMDA receptors by depolarizing the postsynaptic neuronal membrane with release of the Mg^{2+} block. Since ammonia (NH_4^+) and K^+ have similar radii, NH_4^+ can pass through K^+ channels and thus cause depolarization of the postsynaptic neuronal membrane (Binstock and Lecar, 1969). Tsui et al. (2004) considered the possibility that ammonia tolerance in the Oriental weatherloach may be due to the presence in this fish of highly selective K^+ channels which were largely impermeable to NH_4^+. They present evidence in favor of this hypothesis. The presence of highly selective K^+ channels which are relatively impermeable to NH_4^+ in this fish would imply that it would be difficult for ammonia to depolarize the post-synaptic membrane and hence activate NMDA receptors. This conclusion is somewhat at odds with the observation that MK-801 is protective against ammonia toxicity in this fish indicating that NMDA receptors are activated. Perhaps the mechanism by which ammonia activates NMDA receptors in this fish differs from that in mammals.

In summary, the data suggest that activation of NMDA receptors is not a general feature of acute ammonia toxicity in fish and in fact may be species dependent. In addition there is some evidence to suggest that when NMDA receptors are activated in the setting of acute ammonia toxicity in fish, the mechanism differs from the mechanism by which ammonia activates NMDA receptors in mammals.

Summary

In general, fish tolerate higher plasma and brain ammonia levels than mammals. Within the fish family, ammonia tolerance is not uniform and some species such as mudskippers are able to withstand exceptionally high brain ammonia concentrations. Fish have evolved strategies to reduce or avoid ammonia toxicity, but in addition the central nervous system of the fish is intrinsically more resistant to the toxic effects of excess ammonia than mammals. In both mammals and fish, excess ammonia in the brain is detoxified through the synthesis of glutamine from glutamate and ammonia. In fish this represents a very important defense mechanism. However in mammals, there is evidence that high brain glutamine levels can be pathogenic and glutamine has been implicated as one of the factors responsible for brain edema in acute ammonia toxicity. In fish, brain edema does not occur in the setting of acute ammonia toxicity, although this conclusion is based upon studies in the gulf toadfish only. In fish, the data indicate that the accumulation of glutamine in the brain in acute hyperammonemic states is non-pathogenic. In mammals, acute ammonia toxicity appears to be mediated chiefly via the activation of NMDA receptors. Activation of these receptors results in a cascade of events which lead to the toxic manifestations of acute hyperammonemia. The details involved in this scenario are still not well understood. In fish, the role played by activation of NMDA receptors in the toxic manifestations of excess ammonia is unclear. The data are very limited and ammonia induced activation of NMDA receptors may be species dependent. In mammals, ammonia activates NMDA receptors by depolarizing the postsynaptic neuronal membrane. In fish, activation of

NMDA receptors by ammonia may be by a different mechanism than that occurring in mammals.

The central nervous system of the fish has clearly evolved adaptations which enable this vertebrate to tolerate very high levels of ammonia. The current experimental data just give inklings as to the nature of these adaptations. Some of these potential adaptations are as follows. (1) Acute ammonia toxicity does not lead to brain edema in the fish even though acute ammonia toxicity does result in the accumulation of glutamine in the brain. Perhaps acute ammonia toxicity in the fish is not associated with an increase in expression of brain cell membrane water channels as has been demonstrated in mammals. (2) Fish may possess neuronal K^+ channels with a very high selectivity thereby limiting the ability of ammonia to depolarize the neuronal membrane. This adaptation could prevent or at least limit activation of NMDA receptors by ammonia. (3) The role of the glutamatergic synapse in mediating the toxic effects of ammonia in the fish has largely been unexplored. Is activation of NMDA receptors by ammonia species dependent? In those fish in which ammonia does activate NMDA receptors, what is the underlying mechanism; depolarization of the postsynaptic membrane or some other process?

In mammals, the two major factors believed to be responsible for acute ammonia toxicity are accumulation of glutamine in the brain and activation of NMDA receptors. In fish, the experimental evidence suggests that high brain glutamine levels are non-pathogenic. In those species of fish in which NMDA receptors are not activated, what then are the factors responsible for the toxic manifestations of excess ammonia? Currently there are no data as to what these factors might be.

In summary, the fish brain is tolerant of high levels of ammonia although this tolerance varies among species. The major mechanisms responsible for acute ammonia toxicity in mammals do not appear to be operative in fish although only a few species have been studied. Excess ammonia does not cause brain edema or changes in brain energy metabolism in the gulf toadfish. High brain glutamine levels are not pathogenic in fish. In two species of mudskippers, excess ammonia does not activate NMDA receptors. These observations lead to a number of potential research questions. (1) How widespread is ammonia induced activation of NMDA receptors among different fish species? What features of the different species correlate with the presence or absence of NMDA receptor activation by ammonia? (2) In those species of fish in which ammonia does activate NMDA receptors, what is the underlying process responsible for activation? (3) In those species of fish in which ammonia does not activate NMDA receptors, what molecular characteristics of the receptor might account for this "resistance"? (4) Why do high brain ammonia levels in the gulf toadfish not result in brain edema? Is this "resistance" to ammonia induced brain edema species dependent? (5) Why does ammonia not cause changes in brain energy metabolism in the gulf toadfish? Are specific enzymes of the tricarboxylic acid cycle not inhibited by ammonia in this fish as they are in mammals? (6) The mechanisms underlying the neurotoxic effects of ammonia and the neurotoxic effects of hypoxia/ischemia share many similarities. Chapter 6 of this book, contains a section comparing the effects of ammonia and hypoxia/ischemia on neurons. Given this similarity between the neurotoxicity of ammonia and hypoxia/ischemia, are ammonia tolerant fish also hypoxia tolerant? There are no known published studies addressing this question. However, pursuing the answer to this question could be a very fruitful research area.

The fish represents a wonderful natural animal model for studying the problem of ammonia toxicity. Studies in the fish would clearly complement those being performed in mammals. In mammals the basic question being asked is how does excess ammonia cause central nervous system toxicity, whereas in the fish the basic question would be how does the central nervous system of this vertebrate "resist" the toxic manifestations of excess ammonia.

References

Albrecht, J., Dolinska, M., 2001. Glutamine as a pathogenic factor in hepatic encephalopathy. *J. Neurosci. Res.* 65, 1–5.

Anderson, P.M., 2001. Urea and glutamine synthesis: environmental influences on nitrogen excretion. In: Wright, P.A., Anderson, P.M. (Eds.), *Fish Physiology, Volume 20, Nitrogen Excretion*. Academic Press. San Diego, California, pp. 239–277.

Backus, K.H., Kettenmann, H., Schachner, M., 1989. Pharmacological characterization of the glutamate receptor in cultured astrocytes. *J. Neurosci. Res.* 22, 274–282.

Ballantyne, J.S., 1997. Jaws: the inside story. The metabolism of elasmobranch fishes. *Comp. Biochem. Physiol.* 118B, 703–742.

Ballantyne, J.S., 2001. Amino acid metabolism. In: Wright, P.A., Anderson,

P.M. (Ed.), *Fish Physiology, Volume 20, Nitrogen Excretion*. Academic Press. San Diego, California, pp 77–107.

Binstock, L., Lecar, H., 1969. Ammonium ion currents in the squid giant axon. *J. Gen. Physiol.* 53, 342–361.

Bridges, R.J., Hatalski, C.G., Shim, S.N., Cummings, B.J., Vijayan, V., Kundi, A., Cotman, C.W., 1992. Gliotoxic actions of excitatory amino acids. *Neuropharmacology* 31, 899–907.

Brusilow, S.W., Maestri, N.E., 1996. Urea cycle disorders: diagnosis, pathophysiology, and therapy. *Adv. Pediatr.* 43, 127–170.

Buchanan, J.T., Brodin, L., Dale, N., Grillner, S., 1987. Reticulospinal neurons activate excitatory amino acid receptors. *Brain Res.* 408, 321–325.

Butterworth, R.F., 2001. Glutamate transporter and receptor function in disorders of ammonia metabolism. *Men. Retard. Dev. Disabil. Res. Rev. 7*, 276–279.

Butterworth, R.F., 2002. Glutamate transporters in hyperammonemia. *Neurochem. Int.* 41, 81–85.

Campbell, J.W., 1991. Excretory nitrogen metabolism. In: Prosser, C.L. (Ed.), *Environmental and Metabolic Animal Physiology*. Wiley-Liss, New York, pp. 277–324.

Chamuleau, R.A.F.M., Vogels, B.A.P.M., Bosman, D.K., Bovee, W.M.M.J., 1994. *In vivo* brain magnetic resonance imaging (MRI) and magnetic resonance spectroscopy (MRS) in hepatic encephalopathy. In: Felipo, V., Grisolia, S. (Eds.), *Hepatic Encephalopathy, Hyperammonemia, and Ammonia Toxicity*. Plenum Press, New York, pp. 23–31.

Chan, P.H., Chu, L., Chen, S., 1990. Effects of MK-801 on glutamate-induced swelling of astrocytes in primary cell culture. *J. Neurosci. Res.* 25, 87–93.

Chan, P.H., Butterworth, R.F., 2003. Cell-selective effects of ammonia on glutamate transporter and receptor function in the mammalian brain. *Neurochem. Int.* 43, 525–532.

Clemmesen, J.O., Larsen, F.S., Kondrup, J., Hansen, B.A., Ott, P., 1999. Cerebral herniation in patients with acute liver failure is correlated with arterial ammonia concentration. *Hepatology* 29, 648–653.

Clemmesen, J.O., Kondrup, J., Ott, P., 2000. Splanchnic and leg exchange of amino acids and ammonia in acute liver failure. *Gastroenterology* 118, 1131–1139.

Cooper, A.J.L., Plum, F., 1987. Biochemistry and physiology of brain ammonia. *Physiol. Rev.* 67, 440–519.

Cooper, A.J.L., 2001. Role of glutamine in cerebral nitrogen metabolism and ammonia neurotoxicity. *Ment. Retard. Dev. Disabil. Res. Rev.* 7, 280–286.

Cox, J.A., Kucenas, S., Voigt, M.M., 2005. Molecular characterization and embryonic expression of the family of N-methyl-D-aspartate receptor subunit genes in the zebrafish. *Dev. Dyn.* 234, 756–766.

Dantzler, W.H., 1989. *Comparative Physiology of the Vertebrate Kidney.* Springer-Verlag, Heidelberg, Germany.

Deferrari, G., Garibotto, G., Robaudo, C., Ghiggeri, G.M., Tizianello, A., 1981. Brain metabolism of amino acids and ammonia in patients with chronic renal insufficiency. *Kidney Int.* 20, 505–510.

Dejong, C.H.C., Deutz, N.E.P., Soeters, P.B., 1993. Cerebral cortex ammonia and glutamine metabolism in two rat models of chronic liver insufficiency-induced hyperammonemia: influence of pair-feeding. *J. Neurochem.* 60, 1047–1057.

Desjardins, P., Michalak, A., Therrien, G., Chatauret, N., Butterworth, R.F., 2001. Increased expression of the astrocytic/endothelial cell glucose transporter (and water channel) protein Glut 1 in relation to brain glucose metabolism and edema in acute liver failure. *Hepatology* 34, 237A (abstract #254).

Dunn, R.J., Bottai, D., Maler, L., 1999. Molecular biology of the Apteronotus NMDA receptor NR1 subunit. *J. Exp. Biol.* 202, 1319–1326.

Felipo, V., Butterworth, R.F., 2002a. Neurobiology of ammonia. *Prog. Neurobiol.* 67, 259–279.

Felipo, V., Butterworth, R.F., 2002b. Mitochondrial dysfunction in acute hyperammonemia. *Neurochem. Int.* 40, 487–491.

Fine, A., 1982. Effects of acute metabolic acidoisis on renal, gut, liver, and muscle metabolism of glutamine and ammonia in the dog. *Kidney Int.* 21, 439–444.

Harvey-Girard, E., Dunn, R.J., 2003. Excitatory amino acid receptors of the electrosensory system: the NR1/NR2B N-methyl-D-aspartate receptor. *J. Neurophysiol.* 89, 822–832.

Hermenegildo, C., Marcaida, G., Montoliu, C., Grisolia, S., Minana, M.-D., Felipo, V., 1996. NMDA receptor antagonists prevent acute ammonia toxicity in mice. *Neurochem. Res.* 21, 1237–1244.

Hermenegildo, C., Monfort, P., Felipo, V., 2000. Activation of N-methyl-D-aspartate receptors in rat brain *in vivo* following acute ammonia intoxication: characterization by *in vivo* brain microdialysis. *Hepatology* 31, 709–715.

Huizenga, J.R., Gips, C.H., Tangerman, A., 1996. The contribution of various organs to ammonia formation: a review of factors determining the arterial ammonia concentration. *Ann. Clin. Biochem.* 33, 23–30.

Ip, Y.K., Chew, S.F., Randall, D.J., 2001. Ammonia toxicity, tolerance, and excretion. In: Wright, P.A., Anderson, P.M. (Eds.), *Fish Physiology, Volume 20, Nitrogen Excretion.* Academic Press, San Diego, California, pp. 109–148.

Ip, Y.K., Chew. S.F., Randall, D.J., 2004a. Five tropical air-breathing fishes, six different strategies to defend against ammonia toxicity on land. *Physiol. Biochem. Zool.* 77, 768–782.

Ip, Y.K., Tay, A.S.L., Lee, K.H., Chew. S.F., 2004b Strategies for surviving high concentrations of environmental ammonia in the swamp eel *Monopterus albus. Physiol. Biochem. Zool.* 77, 390–405.

Ip, Y.K., Leong, M.W.F., Sim, M.Y., Goh, G.S., Wong, W.P., Chew, S.F., 2005. Chronic and acute ammonia toxicity in mudskippers, *Periophthalmodon schlosseri* and *Boleophthalmus boddaerti*: brain ammonia and glutamine contents, and effects of methionine sulfoximine and MK801. *J. Exp. Biol.* 208, 1993–2004.

Kosenko, E., Kaminsky, Y., Grau, E., Minana, M.-D., Marcaida, G., Grislia, S., Felipo, V., 1994. Brain ATP depletion induced by acute ammonia intoxication in rats is mediated by activation of the NMDA receptor and Na^+, K^+-ATPase. *J. Neurochem.* 63, 2172–2178.

Lockwood, A.H., McDonald, J.M., Reiman, R.E., Gelbard, A.S., Laughlin, J.S., Duffy, T.E., Plum, F., 1979. The dynamics of ammonia metabolism in man. *J. Clin. Invest.* 63, 449–460.

Mans, A.M., DeJoseph, M.R., Hawkins, R.A., 1994. Metabolic abnormalities and grade of encephalopathy in acute hepatic failure. *J. Neurochem.* 63, 1829–1838.

Margulies, J.E., Thompson, R.C., Demtriou, A.A., 1999. Aquaporin-4 water channel is up-regulated in the brain in fulminant hepatic failure. *Hepatology* 30, 395A (abstract #938).

Mathias, R.S., Kostiner, D., Packman, S., 2001. Hyperammonemia in urea cycle disorders: role of the nephrologist. *Am. J. Kid. Dis.* 37, 1069–1080.

Monfort, P., Montoliu, C., Hermenegildo, C., Munoz, M.-D., Felipo, V., 2000. Differential effects of acute and chronic hyperammonemia on signal transduction pathways associated to NMDA receptors. *Neurochem. Int.* 37, 249–253.

Monfort, P., Kosenko, E., Erceg, S., Canales, J.-J., Felipo, V., 2002. Molecular mechanism of acute ammonia toxicity: role of NMDA receptors. *Neurochem. Int.* 41, 95–102.

Olde Damink, S.W.M., Deutz, N.E.P., Dejong, C.H.C., Soeters, P.B., Jalan, R., 2002. Interogan ammonia metabolism in liver failure. *Neurochem. Int.* 41, 177–188.

Porro, G.B., Maiolo, A.T., 1970. Cerebral ammonia metabolism in uremia. *Life Sci.* 9, 43–50.

Purves, D., Augustine, G.J., Fitzpatrick, D., Hall, W.C., LaMantia, A.-S., McNamara, J.O., Williams, S.M., 2004. *Neuroscience,* 3rd Edn. Sinauer Associates, Inc., Sunderland, Massachusetts.

Randall, D.J., Wright, P.A., 1987. Ammonia distribution and excretion in fish. *Fish Physiol. Biochem.* 3, 107–120.

Randall, D.J., Tsui, T.K.N., 2002. Ammonia toxicity in fish. *Mar. Pollution Bull.* 45, 17–23.

Richards, B., Velasques, A., Eiseman, B., 1971. Effect of elevating tissue ammonia concentration on renal function in dogs. *Br. J. Urol.* 43, 536–539.

Rodwell, V.W., 2000. Catabolism of proteins and of amino acid nitrogen. In: Murray, R.K., Granner, D.K., Mayes, P.A., Rodwell, V.W. (Eds.), *Harper's Biochemistry, 25th Edn.* Appleton and Lange, Stamford, Connecticut, pp. 313–322.

Schofield, J.P., Cox, T.M., Caskey, T., Wakamiya, M., 1999. Mice deficient in the urea-cycle enzyme, carbamoyl phosphate synthetase 1, die during the early neonatal period from hyperammonemia. *Hepatology* 29, 181–185.

Shupliakov, O., Pieribone, V.A., Gad, H., Brodin, L., 1996. Presynaptic mechanisms in central synaptic transmission: "biochemistry" of an intact glutamatergic synapse. *Acta. Physiol. Scand.* 157, 369–379.

Shupliakov, O., Ottersen, O.P., Storm-Mathisen, J., Brodin, L., 1997. Glial and neuronal glutamine pools at glutamatergic synapses with distinct properties. *Neuroscience* 77, 1201–1212.

Singer, M.A., 2003. Dietary protein-induced changes in excretory function: a general animal design feature. *Comp. Biochem. Physiol. Part B* 136, 785–801.

Soengas, J.L., Aldegunde, M., 2002. Energy metabolism of fish brain. *Comp. Biochem. Physiol. Part B* 131, 271–296.

Stanley, C.A., 2004. Hyperinsulinism/hyperammonemia syndrome: insights into the regulatory role of glutamate dehydrogenase in ammonia metabolism. *Mol. Genet Metab* 81, S45–S51.

Swain, M., Butterworth, R.F., Blei, A.T., 1992. Ammonia and related amino acids in the pathogenesis of brain edema in acute ischemic liver failure in rats. *Hepatology* 15, 449–453.

Takahashi, H., Koehler, R.C., Brusilow, S.W., Traystman, R.J., 1991. Inhibition of brain glutamine accumulation prevents cerebral edema in hyperammonemic rats. *Am. J. Physiol.* 261, H825–H829.

Teichberg, V.I., 1991. Glial glutamate receptors: likely actors in brain signaling. *FASEB. J.* 5, 3086–3091.

Tsui, T.K.N., Randall, D.J., Hanson, L., Farrell, A.P., Chew, S.F., Ip, Y.K., 2004. Dogmas and controversies in the handling of nitrogenous wastes: ammonia tolerance in the Oriental weatherloach *Misgurnus anguillicaudatus*. *J. Exp. Biol.* 207, 1977–1983.

van Waarde, A., 1981. Nitrogen metabolism in goldfish, *Carassius auratus* (L.). Activities of transamination reactions, purine nucleotide cycle and glutamate dehydrogenase in goldfish tissues. *Comp. Biochem. Physiol.* 68B, 407–413.

Veauvy, C.M., Wang, Y., Walsh, P.J., Perez-Pinzon, M.A., 2002. Comparison of the effects of ammonia on brain mitochondrial function in rats and gulf toadfish. *Am. J. Physiol. Regul. Integr. Comp. Physiol.* 283, R598–R603.

Veauvy, C.M., McDonald, M.D., Van Audekerke, J., Vanhoutte, G., Van Camp, N., Van der Linden, A., Walsh, P.J., 2005. Ammonia affects brain nitrogen metabolism but not hydration status in gulf toadfish (*Opsanus beta*). *Aquat. Toxicol* 74, 32–46.

Vogels, B.A.P.M., Maas, M.A.W., Daalhuisen, J., Quack, G., Chamuleau, R.A.F.M., 1997. Memantine, a non-competitive NMDA receptor antagonist improves hyperammonemia-induced encephalopathy and acute hepatic encephalopathy in rats. *Hepatology* 25, 820–827.

Wang, Y., Walsh, P.J., 2000. High ammonia tolerance in fishes of the family Batrachoididae (toadfish and midshipmen). *Aqua. Toxicol.* 50, 205–219.

Weber, F.L., Veach, G.L., 1979. The importance of the small intestine in gut ammonium production in the fasting dog. *Gastroenterology* 77, 235–240.

Webster Jr, L.T., Gabuzda, G.J., 1958. Ammonium uptake by the extremities and brain in hepatic coma. *J. Clin. Invest.* 37, 414–424.

Welters, C.F.M., Deutz, N.E.P., Dejong, C.H.C., Soeters, P.B., 1999. Enhanced renal vein ammonia efflux after a protein meal in the pig. *J. Hepatol.* 31, 489–496.

Wicks, B.J., Randall, D.J., 2002. The effect of feeding and fasting on ammonia toxicity in juvenile rainbow trout, *Oncorhynchus mykiss*. *Aquat. Toxicol.* 59, 71–82.

Wilson, R.P., Muhrer, M.E., Bloomfield, R.A., 1968. Comparative ammonia toxicity. *Comp. Biochem. Physiol.* 25, 295–301.

Wilson, R.P., Anderson, R.O., Bloomfield, R.A., 1969. Ammonia toxicity in selected fishes. *Comp. Biochem. Physiol.* 28, 107–118.

Young, V.R., Ajami, A.M., 2001. Glutamine: the emperor or his clothes. *J. Nutr.* 131, 2449S–2459S.

CHAPTER 6

Hypoxia/Ischemia

In higher organisms, respiratory and cardiovascular systems provide and distribute oxygen (O_2) to tissues and cells. Oxygen serves as the terminal electron acceptor during mitochondrial oxidative phosphorylation, which is the major biochemical reaction for capturing energy in the form of adenosine triphosphate (ATP). Since the function of the mammalian brain depends on a continuous supply of O_2 and glucose, brain function quickly deteriorates when the supply of these substrates is interrupted (Hansen, 1985). In this chapter, the effects of hypoxia/ischemia on mammalian brain structure and function and what is known about the adaptations of hypoxia tolerant vertebrates such as high altitude birds and certain species of turtle are considered.

In humans, a variety of lung diseases impair alveolar gas exchange with the result that the pO_2 of arterial blood is reduced below the range of normal. Perhaps the most common of these lung diseases is tobacco-induced emphysema. In this disorder, destruction of lung tissue with significant ventilation/perfusion mismatching is the underlying basis for the reduced arterial blood pO_2. In humans with cerebrovascular disease, vessel occlusion results in ischemia of downstream brain tissue with cellular necrosis if the vascular occlusion is not quickly corrected. However, when considering the pathogenesis of brain dysfunction, one must distinguish between hypoxic and ischemic states since the two are quite different. An

ischemic state is one in which there is impairment of blood supply resulting in reduced delivery of substrates (chiefly O_2 and glucose) and reduced
removal of waste products, while a hypoxic state is one in which there is
reduced delivery of oxygen only (oxygen deprivation) with no interrup-
tion of blood flow (Pearigen *et al.*, 1996; Auer and Sutherland, 2002). In this
chapter the terms hypoxia and ischemia will be used to refer to a hypoxic
and an ischemic state, respectively.

In the next sections, relevant features of mammalian cerebral circulation
and brain energy metabolism are covered, followed by a review of the
pathogenesis of hypoxic and ischemic brain dysfunction in the mammal.

Cerebral Blood Flow: Its Regulation

Cerebral blood flow (CBF) is regulated by both systemic and local factors.
Kety and Schmidt (1948) demonstrated that in young male volunteers CO_2
inhalation (5%–7%) resulted in a 75% increase in CBF and inhalation of 10%
O_2 increased CBF by 35%. Both hypercapnia and hypoxia led to a decrease
in cerebrovascular resistance indicating that the increased CBF was the
result of vasodilatation. These investigators described a curvilinear relationship between arterial pCO_2 and CBF. A reduction of pCO_2 from 40 to
20 mmHg decreased CBF but not to the same extent as the increase in CBF
resulting from the change in pCO_2 from 40 to 60 mmHg. Cerebral oxygen
consumption did not significantly change during the experimental periods
when the volunteers were inhaling inspired air with excess CO_2 (5%–7%)
or low O_2 content (10%). Hence under these circumstances, the increase in
cerebral blood flow without an increase in cerebral oxygen consumption
resulted in a decrease in the arterial-venous oxygen difference (reduced
oxygen extraction). Kety and Schmidt used the nitrous oxide method for
measuring CBF which requires at least ten minutes of steady state circulation and metabolism for each blood flow measurement. Shimojyo *et al.*
(1968) studied the effects of graded hypoxia in humans using a radioactive
isotope method to measure CBF. This method allowed minute to minute
observations of whole CBF. Hypoxia was induced by having the subjects
breathe a gas mixture of 6% O_2-94% N_2. The cerebral rate of oxygen consumption ($CMRO_2$) was measured simultaneously. No effort was made
to control the arterial pCO_2 during the period of hypoxia. There was no
significant change in mean CBF until arterial pO_2 fell below 30 mmHg.
Below this point CBF increased rapidly with further hypoxia even though

pCO_2 was not controlled. The arterial pO_2 range studied was from about 23 mmHg to 100 mmHg. There was no significant change in $CMRO_2$ over this arterial pO_2 range (Shimojyo *et al.*, 1968; Gjedde *et al.*, 2002). Brown *et al.* (1985) measured CBF in normal individuals and in patients with hematologic disorders resulting in hemoglobin concentrations varying from 5.7 to 19.1 g/100 ml. These patients would have different arterial oxygen contents but not different arterial pO_2 tensions. A close inverse relationship was found between arterial oxygen content and CBF. Cerebral oxygen transport, which was calculated as the product of CBF times arterial oxygen content, was constant across the range of hemoglobin concentrations studied. However, the most interesting aspect of CBF regulation is the influence of local brain activity. Local brain activity determines not only oxygen and glucose consumption but also blood flow in any particular brain region (Auer and Sutherland, 2002; Lindauer *et al.*, 2003). For example, Hoge *et al.* (1999) demonstrated the coupling of regional cerebral blood flow and oxygen metabolism in a series of experiments in humans. Blood flow and the cerebral metabolic rate of oxygen consumption ($CMRO_2$) were measured in the primary visual cortex during sustained graded visual stimulation. Fractional changes in blood flow and $CMRO_2$ were linearly coupled in a consistent ratio of approximately 2:1 (blood flow:$CMRO_2$). They also calculated that elevated energy demands during brain activation are met largely through oxidative metabolism. This coupling of neuronal activity to regional CBF involves factors such as nitric oxide, adenosine and potassium (Lindauer *et al.*, 2003).

Lindauer *et al.* (2003) compared the neuronal activated regional CBF response in anesthetized rats under conditions of normoxia and hypoxia. Hypoxia was induced by addition of N_2 to the inspired room air. Mean arterial pO_2s for normoxic and hypoxic states were 130 and 45 mmHg respectively. These investigators measured relative changes in oxygenated (oxy-Hb) and deoxygenated hemoglobin (deoxy-Hb) concentrations using a custom made microfiber spectroscope connected to a spectrometer over the surface area of the stimulated brain region. Somatosensory evoked potentials were simultaneously recorded from the stimulated region. Changes in oxy-Hb and deoxy-Hb concentration were plotted versus time. Activation of a brain region (in this case the whisker barrel cortex was stimulated by deflection of whisker hairs) induces a response consisting of an increase in oxy-Hb concentration and a decrease in deoxy-Hb concentration reflecting the inflow of arteriolarized highly oxygenated blood. In the experiments of Lindauer *et al.* (2003) stimulation time was four seconds. This hyperoxygenation response occurred one to two seconds after stimulation onset and peaked at about three to four seconds. Hemoglobin oxygenation changes returned to baseline within ten seconds. During the state of hypoxia, relative peak responses of the oxy-Hb increase and the deoxy-Hb decrease were greater than during normoxia (mean changes; oxy-Hb increase 145% of normoxic response, deoxy-Hb decrease 152% of normoxic response). These observations indicate that neuronal activity results in an increase in regional cerebral blood flow with a flooding of highly oxygenated blood into the activated area. In addition, these experimental results demonstrate that changes in systemic blood oxygenation significantly influence regional cerebral blood oxygenation changes during neuronal functional activation. Low systemic arterial pO_2 increases the hyperoxgenation response to somatosensory stimulation in the rat. This observation implies that the signals initiating increases in blood flow are additive. In this case the increase in flow induced by hypoxia is added on to the increase in flow associated with elevated levels of neural activity. Hence, regional cerebral blood flow is regulated according to the metabolic demand of oxygen within the activated tissue. The signals initiating the increase in regional blood flow in functionally active brain areas are unknown but Lindauer *et al.* (2003) have concluded that tissue hypoxia is unlikely to be the initiating signal at least in the anesthetized rat. Jones *et al.* (2001) studied regional blood flow and oxygen consumption in the rodent barrel cortex under two sets of circumstances. Electrical stimulation of the whisker pad (at varying intensities) was used to activate this region and increasing the fraction of inspired pCO_2 ($FICO_2$) was used to induce a hemodynamic response. Neural activation resulted in an increase in both fractional regional blood flow and fractional concentration of oxygenated hemoglobin and a decrease in fractional concentration of deoxygenated hemoglobin, similar to the observations of Lindauer *et al.* (2003). In addition, increases in neural activity were accompanied by increases in oxygen consumption even at the lowest stimulation intensity. The ratio of changes in regional blood flow to changes in regional oxygen consumption was similar to that obtained by Hoge *et al.* (1999). The hypercapnic state (an $FICO_2$ of 4% increased arterial pCO_2 by about 19.5 mmHg) was associated with an increase in regional blood flow. However, somewhat surprisingly, the increased blood flow induced by hypercapnia did result in an increase in oxygen consumption although for any given blood flow value the increase in oxygen consumption following neural activation was well above that observed with hypercapnia. This observation is at odds with the data of

Kety and Schmidt (1948) who observed no change in global cerebral oxygen consumption with hypercapnia.

Tissue oxygen tension (ptO_2) has been measured in the rat cerebral cortex by Sick *et al.* (1982a) using an O_2-sensitive microelectrode. A histogram of ptO_2 values was obtained by making multiple cortical penetrations of $100\,\mu$m steps with the microelectrode. During normoxia (arterial pO_2 > 110 mmHg) the distribution of ptO_2 values was right skewed with values as low as 1–2 mmHg and an occasional value as high as 45 mmHg. The peak frequency occurred near 10 mmHg. Thus under normal conditions tissue oxygen tension is low suggesting that the mammalian brain can tolerate low oxygen environments. The low tissue oxygen level implies that the rate of cellular oxygen metabolism is limited by the rate of diffusion of oxygen from capillary to brain cell. In fact the data of Hoge *et al.* (1999) already discussed is consistent with this general notion of diffusion-limited oxygen delivery.

The movement of oxygen from capillary to parenchymal cell mitochondria is a very complex process but there have been a number of formal mathematical models [see for example Groebe and Thews (1992) and Buxton and Frank (1997)]. The following is a very simplistic description of this process. Most of the oxygen (about 98%) transported in the blood is bound to hemoglobin within erythrocytes. The exchange of oxygen between the pool of oxygen dissolved in the plasma and oxygen bound to hemoglobin in the erythrocytes is very rapid (Buxton and Frank, 1997; Groebe and Thews, 1992). The driving force for the flux of oxygen between capillary and mitochondria is the pO_2 gradient. Each oxygen molecule in plasma has a probability per unit time of moving out of the capillary, i.e. being extracted. Since there is a pO_2 (or oxygen concentration) gradient across the capillary wall, applying this probability to all of the oxygen molecules in the plasma will give the net flux of oxygen out of the capillary. The oxygen extraction fraction (OEF) can be defined as the net flux of oxygen out of the capillary per unit time divided by the total number of oxygen molecules passing through the capillary in that same unit of time. In humans OEF is about 30%–55% (Buxton and Frank, 1997). OEF bears an inverse nonlinear relationship to the rate of blood flow (Buxton and Frank, 1997). An increase in rate of blood flow will result in a decrease in OEF of lesser magnitude than that of the increase in flow. Hence an increase in capillary blood flow will result in more oxygen molecules passing through the capillary per unit time and a greater number of oxygen molecules moving out of the capillary per unit time, since OEF falls to a lesser extent than the increase in flow rate. In support of this model, a plot of OEF versus percentage increase in cerebral blood flow using the data of Kety and Schmidt (1948) gives a curve similar to that presented by Buxton and Frank (1997). In addition, the reduction in OEF will raise average capillary pO_2 since there will be smaller decrease in pO_2 as the blood travels down the capillary. The increase in average capillary pO_2 will lead to an increase in tissue pO_2. As expected then based on this simple model, the neuronal-activity induced increase in regional blood flow leads to an increase in capillary and tissue pO_2 as pointed out by Buxton (2001) and Gjedde *et al.* (2002). An increase in hemoglobin concentration will increase the number of oxygen molecules available to cross the capillary wall since the pools of oxygen dissolved in the plasma and oxygen bound to hemoglobin are in equilibrium (Buxton and Frank, 1997; Groebe and Thews, 1992). Hence for a given OEF, there will be a larger flux of oxygen molecules across the capillary wall when the hemoglobin concentration is raised.

Brain tissue oxygen tension will reflect the balance between oxygen delivery and oxygen consumption as determined by neuronal and glial activity. This topic is discussed in more detail by Gjedde *et al.* (2002).

Brain Energy Metabolism

Brain tissue represents 0.1%–1% of the body weight of vertebrates (excluding primates) but is responsible for 1.5%–8.5% of total body energy consumption in endothermic vertebrates and a comparable range (2.7%–3.4%) is found for ectothermic vertebrates (Soengas and Aldegunde, 2002). Unlike other organs, the brain generally uses only glucose as a fuel and at steady state, oxidative metabolism accounts for 90% of the breakdown of glucose (Gjedde *et al.*, 2002). Most of brain energy consumption is used to maintain ionic gradients across plasma membranes and to restore these gradients after depolarization. Approximately 70%–80% of total ATP turnover is consumed by ion transport pumps, with just one such pump, the Na^+-K^+-ATPase pump accounting for at least 50% of ATP turnover in mammalian cells (Bickler and Buck, 1998). What is still unclear is how the metabolic processes involved in ATP production are partitioned among the different cells in the brain. Recently it has been proposed that during neuronal excitation there is net generation of lactate by astrocytes [which possess the glycolytic lactate dehydrogenase (LDH) subtype LD5] with net transfer of lactate from astrocytes to neurons via monocarboxylate transporters.

Within the neurons lactate is converted to pyruvate (via the oxidative LDH subtype, LD1) which then enters the tricarboxylic acid (TCA) cycle (Gjedde and Marrett, 2001). This hypothesis was tested by Gjedde and Marrett (2001) who measured $CMRO_2$ for the visual cortex in human volunteers during sustained stimulation with a complex stimulus. These investigators reported that oxidative metabolism ($CMRO_2$) of visual cortex increased significantly with continuous stimulation (up to 11 minutes) and that the additional supply of pyruvate accompanying the rise of oxygen consumption originated in a tissue compartment with a preponderance of the LDH subtype LD1, i.e. the neuron. The results of this study indicate that during functional activation, neurons increase their oxidative metabolism in proportion to an increase in pyruvate levels generated by neuronal rather than astrocytic glycolysis.

The topic of oxidative and non-oxidative metabolism in excited neurons and astrocytes of the mammalian brain has recently been reviewed by Gjedde *et al.* (2002). These authors point out that the complexity of the stimulus is an important determinant of subsequent changes in oxygen consumption and ATP production in the excited region of the brain. So called simple stimuli such as somatosensory and visual stimuli with little information content, produce a small increase in oxygen consumption and almost no increase in ATP production. On the other hand, complex somatosensory stimuli and complex visual stimuli (much more informational content) are associated with significant increases in oxygen consumption, rate of glycolysis and ATP production. With both types of stimuli there were reasonably comparable increases in regional cerebral blood flow. The observation that simple stimuli produce an increase in local blood flow but very little increase in oxygen consumption appears inconsistent with the concept that diffusion-limited oxygen delivery is the determinant of the rate of oxygen metabolism as discussed in the previous section. The increase in blood flow would lead to an increase in the net flux of oxygen across the capillary wall and an increase in tissue pO_2 as discussed in the previous section. If the rate of oxygen delivery was the limiting factor than oxygen metabolism should increase significantly with simple stimuli.

The difference in the effects of the two types of stimuli could be due to the extent of postprocessing that the stimuli elicit. Simple stimuli may induce little increase in oxidative metabolism because these stimuli result in only limited postsynaptic depolarization with failure to activate the mitochondria concentrated in dendrites. The authors (Gjedde *et al.*, 2002) point out that astrocytes are now believed to carry out more oxidative metabolism than previously believed and that almost half of their glucose supply is oxidized to carbon dioxide. The authors suggest that there are three different levels of focal activity in the brain. The first level is characterized by baseline levels of blood flow and oxygen metabolism. The second level ("stand-by") is characterized by increased blood flow but little increase in oxygen consumption because of inhibition of postsynaptic depolarization. A third level is characterized by increased blood flow and increased oxygen consumption because of significant postsynaptic depolarization and efferent projection from the site. The "stand-by" level is observed with simple stimuli whereas the fully activated level is seen with complex stimuli. The afferent (preprocessing) activity involves primarily astrocytes and the presynaptic terminals whereas the efferent activity involves the dendrites (and perhaps cell bodies) of projection neurons and interneurons. Gjedde *et al.*, (2002) have attempted to quantitate the contributions of astrocytes and neurons *in vivo* to the metabolic changes associated with the proposed different levels of activation. Regional blood flow and astrocytic consumption of glucose and oxygen increase significantly above baseline values when the proposed "stand-by" level of activity is achieved. Blood flow and glucose/oxygen consumption by astrocytes show little further increase when activation moves from "stand-by" to the third level of full activation. Conversely, oxygen and glucose consumption by neurons increase only modestly above baseline values with "stand-by" activation but increase significantly when neurons pass from "stand-by" to full activation. The data presented by Gjedde *et al.* (2002) in Table 2 of their paper, suggest that in neurons *in vivo*, glucose and oxygen consumption are matched such that there is no net production of lactate and pyruvate and hence neurons use no lactate or pyruvate from astrocytes (Gjedde and Marrett, 2001). In astrocytes, these two fluxes are not matched so nét production of lactate and pyruvate does occur. Neurons occupy 50% and astrocytes 33% of the volume of the cerebral cortex *in vivo*. However, again using the data in Table 2 of the Gjedde *et al.* paper, astrocytes account for 50% of the total increase in glucose consumption but only about 15% of the total increase in oxygen consumption and ATP production when a region of the brain goes from baseline status to full activation. The "stand-by" condition is established by afferent activity with release of excitatory neurotransmitters from presynaptic neurons. However, the postsynaptic depolarization is prevented by inhibition. The oxidative response is low because the mitochondria located in dendrites are not stimulated. When the inhibition is lifted, postsynaptic depolarization occurs and a significant increase in oxygen consumption takes place. The changes in metabolism occurring in each level appear to be additive with no significant transfer of lactate from astrocytes to neurons. Also there is apparently no evidence that glycolysis in astrocytes supports oxidative metabolism in neurons in the baseline state. According to this model, in addition to the release of neurotransmitters from the presynaptic terminals and the import of neurotransmitters by astrocytes, a major function of the afferent activity is to establish a larger regional blood flow sufficient to support the increased level of oxidative metabolism occurring in the fully activated state when postsynaptic depolarization ensues and mitochondria are activated in dendrites. The increased blood flow rate overcomes diffusion limitations in oxygen delivery. In addition, the greater rate of neuronal glycolysis in the fully activated state compared to the "stand-by" condition provides the increase in neuronal cell pyruvate necessary for the increased rate of oxidative metabolism.

Before proceeding with a discussion of hypoxic and ischemic brain damage, the salient observations with respect to cerebral blood flow (CBF) and brain energy metabolism are summarized as follows.

(1) Hypoxia induces an increase in global CBF although this increase is only observed when arterial pO_2 falls below a threshold value of 30 mmHg. Even in the presence of an arterial pO_2 as low as 23 mmHg, there is no change in global cerebral metabolism of oxygen ($CMRO_2$).

(2) Activation of a local area of the brain induces a local hemodynamic response consisting of an increase in both local blood flow and the concentration of oxygenated hemoglobin and a decrease in the concentration of deoxygenated hemoglobin. The signals initiating this hemodynamic response are unknown but appear to be additive since systemic hypoxia and neural activation together induce a greater hemodynamic response than either alone.

(3) Brain tissue oxygen tension is low but increases as a result of the local hemodynamic response. This increase results in an enhanced rate of oxygen delivery to brain parenchymal cells.

(4) Oxidative metabolism is a vital feature of neurons and to a lesser extent astrocytes. Full activation of a local brain region is associated with about a 25% increase in total ATP production with about 85% of this amount due to neuronal oxidative metabolism. Neurons appear to match oxygen and glucose consumption *in vivo* with little net production of lactate and pyruvate.

(5) The local hemodynamic response appears designed to deliver sufficient oxygen (and presumably glucose) to allow the required increase in oxidative metabolism for full neuronal activation.

Oxygen and Oxygen/Glucose Deprivation and the Brain

General features of neuronal response

In this section, studies in which the investigators have attempted to separate the effects of hypoxia (oxygen deprivation) and combined oxygen/glucose deprivation on brain function are reviewed. As noted in the introduction, deprivation of both oxygen and glucose would correspond more closely to what has been termed an ischemic state. Comparing the results of different studies is in general not an easy task. In the mammal all neurons are not equally anoxia sensitive. In addition, experimental designs vary. For example, anoxia (oxygen deprivation) is generally induced by replacing inspired oxygen with nitrogen. However, sometimes chemical anoxia is produced by using cyanide to poison the mitochondrial respiratory chain. Cyanide poisoning and oxygen deprivation may not give comparable results (Ballanyi and Kulik, 1998). In addition, tissue oxygen levels are often not well documented and hence it becomes problematic comparing the severity of the anoxic insult between different studies. This latter factor is probably the one rendering the comparison of different experimental results most difficult. Since tissue oxygen levels are not always well documented, the terms hypoxia and anoxia will be used interchangeably. However, in general hypoxia or anoxia will refer to a state of severe oxygen deprivation. In many of the studies, interstitial (extracellular) concentrations of ions are measured and changes in these concentrations are used as a measure of changes in cellular release or uptake of these particular ions. As will become evident, the critical early events induced by anoxia or ischemia are the dissipation of ion gradients and the resultant membrane depolarization. These events trigger a cascade of effects leading to cell death. Those studies in which the investigators have examined these early events will be the main focus.

Pearigen *et al.* (1996) studied rats made hypoxic by replacing the inspired air mixture of 24% oxygen, 76% nitrogen by a mixture of 6% oxygen, 94% nitrogen for 20 minutes. In these animals mean arterial pO_2 fell from a baseline of 87.6 mmHg to 20.4 mmHg five minutes after the onset of hypoxia. Mean arterial blood pressure was not changed by the hypoxic insult. In fact these investigators were not able to study more profound levels of hypoxia since a lower arterial pO_2 resulted in a fall in blood pressure. Brain oxygen tension (measured within the hippocampus) fell to a mean value of 32% of baseline by the end of the 20-minute hypoxic period. Absolute values for brain tissue oxygen tension were not reported. The investigators used the technique of *in vivo* microdialysis to measure brain extracellular glutamate concentration and they observed no significant change in extracellular glutamate concentration during the hypoxic period compared to baseline. Seventy-two hours after recovery, the animals were sacrificed and the brains examined for evidence of morphologic injury using staining for heat shock protein (HSP-72) or evidence for neuronal death using acid fuchsin staining. There was no evidence of brain injury or neuronal death in the hypoxic animals. In contrast, animals made globally ischemic by vessel occlusion for 20 minutes showed evidence of neuronal death throughout all hippocampal regions, thalamus and cortex.

Sick *et al.* (1982a) compared brain tissue oxygen tensions (ptO_2) in the rat parietal cortex (an hypoxic sensitive area) and in the telencephalon of freshwater diving turtles (*Pseudeymys scripta elegans*). This turtle can survive submersion in oxygen free water for at least six months. Part of this study has already been described in the "Cerebral Blood Flow: Its Regulation" section. The mean ptO_2 was similar in the rat (12.9 mmHg) and turtle (12.5 mmHg). In the rat, the distribution of ptO_2s was right skewed while in the turtle the distribution was normally distributed. During normoxia (respiration of rats with 30% oxygen and turtles 21% oxygen) mean extracellular brain potassium concentration was 3.62 mM in rat parietal cortex and 2.59 mM in turtle telencephalon (Sick *et al.*, 1982b). Hypoxia was induced by replacing inspired gas with 100% nitrogen. Turtles were maintained at room temperature; rat temperature was maintained at 37°C. In the rat, ptO_2 reached 0 mmHg within two minutes of nitrogen respiration while in the turtle, at least ten minutes of nitrogen respiration was required to reach a ptO_2 of 0 mmHg. In the rat, the extracellular potassium concentration initially increased gradually upon respiration with 100% nitrogen but within three minutes abruptly rose to a mean level of 77.7 mM reflecting a significant efflux of potassium from the cell. In the turtle, the mean extracellular potassium concentration only rose to 3.68 mM even after 30 minutes of nitrogen respiration. There was never an abrupt rise in extracellular potassium concentration as observed in the rat even when nitrogen respiration continued for as long as four hours. The rapid rise in the extracellular concentration of potassium in the rat is associated with depolarization of nerve and probably glial cells (Hansen, 1985). In the turtle, no rapid rise in extracellular potassium concentration occurred and there was no evident change in membrane potential even after several hours of anoxia (Haddad and Jiang, 1993a). In both the rat and turtle, anoxia led to an increase in the ratio of reduced/oxidized cytochrome aa_3 with full reduction of cytochrome aa_3 occurring within the timeframes required for ptO_2 to fall to 0 mmHg (two minutes in the rat, ten minutes in the turtle). Finally, electrocorticographic activity of the turtle brain was only moderately depressed during nitrogen respiration for up to four hours whereas electrocorticographic activity of rat brain (parietal cortex) became isoelectric within one minute of nitrogen respiration. When respiration with control gas mixture was resumed (30% oxygen in rats, 21% oxygen in turtles) after about three and a half minutes in the rat and 30 minutes in the turtle, the following changes occurred in both species: (1) ptO_2 became elevated transiently above baseline values, cytochrome aa_3 became transiently more oxidized than resting levels and then both returned to pre-anoxic levels; (2) the elevated extracellular potassium concentration fell transiently below the resting level and then returned to baseline value; and (3) complete recovery of electrocorticographic activity.

These studies by Pearigen *et al.* (1996) and Sick *et al.* (1982a and b) indicate that short term hypoxia/anoxia in the mammal causes a loss of ion gradients, membrane depolarization, loss of cortical electrical activity and inhibition of oxidative metabolism. However, there were no morphological evidence of injury or cell death and the functional changes described by Sick *et al.* (1982b) were reversed when normal oxygenation was resumed.

Hansen (1985) reviewed the effects of anoxia on ion distribution in the brain. Most of the data were obtained from experiments using ion sensitive microelectrodes. In the majority of these studies interstitial ion concentrations were measured with the inference being that changes in these interstitial concentrations reflect changes in cellular uptake or release of these ions. Hansen does not always clearly distinguish between anoxia (oxygen deprivation) and ischemia (oxygen/glucose deprivation). For example Fig. 1 of his article summarizes changes in interstitial potassium ion concentrations and electrical activity in the brains of rats following cardiac arrest; clearly an ischemic state. Yet in the text Hansen refers to these changes as occurring during anoxia. In the normal mammal, interstitial ion concentrations were: potassium between 2.8 and 4.0 mM, sodium between 133 and 154 mM and calcium 1.2 to 1.3 mM. The interstitial space is confined to the narrow clefts between the densely packed nerve and glial cells and occupies about 20% of brain volume. Normal intracellular ion concentrations using a variety of techniques were: potassium 87 to 163 mM and sodium 23 to 57 mM. Intracellular calcium concentrations listed in this review article are too high. Interstitial ion concentrations change as a result of neuronal activity. The potassium concentration may increase to 10–12 mM and the calcium concentration may decrease to 0.5 mM. Changes in interstitial sodium concentrations with activity are small. In the rat made ischemic by induction of cardiac arrest, there is an initial slow increase in interstitial potassium concentration to about 10 mM over two minutes followed by a sudden steep rise to about 60 mM over a few seconds. It is interesting that the electroencephalogram becomes isoelectric within 30 seconds following the ischemic insult, when the interstitial potassium concentration is still within the normal range. The rapid rise in the interstitial concentration of potassium (due to efflux of potassium from the cell) is accompanied by significant decreases in the interstitial concentrations of sodium, chloride and calcium (due to influx into the cell). Depolarization of neurons and glial cells (inferred by the changes measured in the DC potential and not by direct measurement of cell membrane potentials) accompanies the dissipation of these ion gradients. Interstitial pH also decreases during ischemia reflecting the anaerobic conversion of glucose to lactic acid. Hansen points out that during anoxia the interstitial space shrinks from about 20% to about 9%–16% of brain volume and brain cells swell. The swelling involves chiefly neurons with very little swelling of glial cells. However it is not clear whether Hansen is referring to oxygen deprivation alone or ischemia during his discussion of the changes in interstitial volume and cell swelling. Such a decrease in the interstitial space would only partially explain the changes in the interstitial ion concentrations with anoxia. As pointed out by Hansen, the rate of anaerobic glycolysis would have to increase by about 15 fold to maintain the preanoxic rate of ATP production; obviously an impossibility. ATP levels decline rapidly in the ischemic brain and the rapid increase in the interstitial concentration of potassium takes place when the ATP level is about 10% of its normal value (Hansen, 1985).

Hypoxia sensitive neurons such as hypoglossal motoneurons in the adult rat undergo a major membrane depolarization within several minutes of oxygen deprivation (Haddad and Jiang, 1993a). This depolarization reaches a maximum change of about 30–40 mV within five to six minutes. Haddad and Jiang (1993b) studied the mechanism underlying this anoxia induced membrane depolarization using *in vitro* brainstem neurons. Anoxia (induced by replacing oxygen in the bathing medium with nitrogen) caused depolarization in hypoglossal neurons and neurons in the dorsal motor nucleus of the vagus (DMX) within three to five minutes. The range of depolarization with anoxia varied from 15–27 mV in DMX and from 34–69 mV in hypoglossal neurons. Extracellular glutamate levels increase during anoxia. However the highest concentration of extracellular glutamate is actually achieved after termination of the anoxic insult (Young *et al.*, 1993). Haddad and Jiang (1993b) demonstrated that blockers of glutamate receptors (NMDA and non-NMDA) had no significant effect on the magnitude or time course of the anoxia induced depolarization in hypoglossal and dorsal vagal motoneurons. (The biology of glutamate receptors is discussed in detail in the "Ammonia Toxicity" chapter 5.) Using hypoglossal neurons, Haddad and Donnelly (1990) showed that the magnitude of the hypoxia induced depolarization was not decreased by tetrodotoxin, calcium free bathing medium or low calcium/high magnesium bathing solutions. Hence, the depolarization did not appear to be due to a tetrodotoxin sensitive sodium influx or to calcium influx. The increase in extracellular potassium as a result of potassium efflux from the cell during hypoxia could account for some of the depolarization. However, Haddad and Jiang (1993b) showed that reducing the concentration of sodium in the extracellular fluid (from 140 to 5 mM) resulted in a significantly smaller depolarization with anoxia (about 18% of the magnitude of the depolarization with hypoxia and normal sodium concentration). Hence, increased entry of sodium (but not through channels sensitive to tetrodotoxin inhibition) into the cell is a major factor in the anoxia induced depolarization. Haddad and Jiang (1993b) also examined the effect of intracellular ATP levels on anoxia induced depolarization. The question under consideration was whether the depolarization was secondary to a decrease in intracellular ATP. In one set of experiments, ATP was iontophoresed intracellularly using conventional microelectrodes. The results of these experiments were mixed with some neurons showing a smaller depolarization after ATP iontophoresis. Because of the possibility that ATP could not be iontophoresed adequately, a second set of experiments were performed in which ATP was present in patch pipettes in order to clamp the concentration of ATP. In these experiments, anoxia induced depolarizations were smaller when ATP was present in the pipette than when ATP was absent.

Hansen (1985) has examined the link between energy metabolism and ion homeostasis in rat brain during ischemia induced by cardiac arrest. The content of glucose in the brain was changed by altering the plasma concentration of this substrate. The time between the onset of the ischemic insult and the rapid increase in extracellular potassium concentration was significantly prolonged in rats made hyperglycemic compared to control normoglycemic rats. Rats made hypoglycemic (with injection of insulin) had a significantly shorter time period between onset of the ischemic insult and the rapid rise in extracellular potassium. These observations suggest that changing brain substrate stores available for immediate anaerobic glycolysis will affect the rate of decline in ATP levels during ischemia and thereby affect the length of time ion gradients can be maintained. These results also suggest that the mechanism underlying the dissipation of ion gradients and resultant membrane depolarization is an energy-linked process of some kind. Hansen (1985) has speculated on how the anoxia induced membrane depolarization with a rapid rise in the extracellular concentration of potassium and a rapid fall in the extracellular sodium concentration is related to depletion of ATP levels and impairment of the Na^+-K^+-ATPase pump. (This pump transports two potassium ions into the cell for every three sodium ions removed from the cell.) He calculated that since the pump counteracts equivalent leaks, even complete failure of the pump with no change in underlying membrane permeability could not account for the experimentally measured rate of increase in the extracellular potassium concentration and rate of decrease in the extracellular sodium concentration that occurs during the anoxia induced depolarization. He proposed that the most likely explanation for the rapid changes in ionic concentrations had to be an increase in the ion permeability of the cell membrane. However it has been well documented that inhibition of the Na^+-K^+-ATPase pump under normoxic conditions does lead to membrane depolarization (Tanaka et al., 1997; Haddad and Donnelly, 1990; Lipton, 1999; Jarvis et al., 2001).

Given that dissipation of ion gradients with the resultant membrane depolarization are the important early events induced by anoxia, it is worthwhile to look at the results of some other studies which shed light on these early events. Using hippocampal brain slices from adult rats, Fried et al. (1995) measured changes in ion transport following a short (five-minute) eriod of anoxia (replacing oxygen in the bathing medium with nitrogen). These investigators recorded postsynaptic action potentials in the CA1 pyramidal region after stimulation of the presynaptic Schaffer collaterals. Anoxia resulted in a loss of the postsynaptic action potential which showed little evidence of recovery even after 60 minutes postanoxic insult. This damage to transmission in hippocampal slice neurons correlated well with damage measured in vivo. At the end of the anoxic period mean intracellular sodium concentration had increased to 187% and mean intracellular potassium concentration had decreased to 70% of the levels found under normal oxygen conditions. The increase in intracellular sodium concentration induced by anoxia was reduced by tetrodotoxin to 119% of control and the anoxia induced decrease in intracellular potassium was of lesser magnitude (89% of control) when tetrodotoxin was present. During ten minutes of anoxia, calcium uptake (measured with radioactive calcium) increased to 113% of the value obtained under normal oxygenation. Mean ATP levels decreased to 49% of normoxic levels after five minutes of anoxia. More recent data with regard to the effects of anoxia on sodium ion fluxes have been reported by Hammarstrom and Gage (2002). The increase in intracellular sodium concentration is due to enhanced influx and not reduced efflux. The increase in intracellular sodium precedes the rise in intracellular free calcium. Reducing the external sodium concentration or application of sodium channel blockers blocks the rise in intracellular calcium. Sodium channel blockers also block the rise in extracellular glutamate concentration and reduce the decrease in ATP levels. Because the sodium influx and depolarization recorded during anoxia are sustained the classical transient sodium channel current could not be responsible. In fact this current is actually reduced during anoxia (Hammarstrom and Gage, 2002; Cummins *et al.*, 1991). A persistent sodium current has been described (Hammarstrom and Gage, 1998 and 2000) in CA1 hippocampal neurons which is resistant to inactivation and which is significantly increased in amplitude during anoxia. Hammarstrom and Gage (1998) have proposed that an increase in this non-inactivating (persistent) sodium current is probably the primary cause of the increase in intracellular sodium observed during anoxia. Using the patch clamp technique, Hammarstrom and Gage (2000) showed that under normal oxygenation the activity (probability of channel openings) of persistent (non-inactivating) sodium channels was very low, but that anoxia significantly increased the activity of these sodium channels. The current-voltage relationship of these channels during normoxia and hypoxia suggested that there was no obvious change in single sodium channel conductance during hypoxia and that these channels were probably also permeable to potassium. Within two to four minutes of anoxia exposure, current amplitude through these channels increased nine fold. Longer periods of anoxia (four to eight minutes) increased current amplitude 18 fold compared to normal oxygen conditions. The activity of these channels could be blocked by tetrodotoxin. If these sodium channels are responsible for the influx of sodium during anoxia and depolarization of the membrane, then the observation that their activity can be blocked by tetrodotoxin is in conflict with the observation of Haddad and Donnelly (1990). As already noted, these investigators reported that in hypoglossal neurons, tetrodotoxin did not reduce the magnitude of the anoxia induced depolarization. The study by Hammarstrom and Gage (2000) demonstrated that persistent sodium channels could sense hypoxia directly since measurements were made in patches excised from hippocampal cells, i.e. in the absence of any cytosolic products of metabolism or second messengers. Hammarstrom and Gage estimated that these channels were responding to a pO_2 of 45 mmHg or less. They proposed that the effects of increased persistent channel activity during anoxia included depolarization of the resting membrane potential, a prolonged and significant influx of sodium into the cell and an increase in the activity of the Na^+-K^+-ATPase pump. Increased activity of this pump would accelerate depletion of ATP stores. Currently it is not known how this sodium channel senses oxygen levels (Hammarstrom and Gage, 2002) although the evidence suggests that the oxygen sensor is attached to the plasma membrane and involves a redox reaction.

It should be noted however, that low oxygen levels do not necessarily lead to an increase in ion channel activity. There are voltage-gated potassium and calcium ion channels that are inhibited by hypoxia (Bickler and Donohoe, 2002). For example, Jiang and Haddad (1994) using the patch clamp technique (cell free excised membrane patches) in substantia nigra and neocortical neurons demonstrated that anoxia reversibly inhibited a class of voltage–gated potassium channels that are inhibited by ATP and activated by calcium. Graded hypoxia was produced by perfusing patches with solutions containing differing oxygen concentrations (2%, 1%, 0%) for at least two hours. Channel activity was almost completely inhibited when pO_2 was less than 1 mmHg and was unaffected when pO_2 was greater than 20 mmHg. The inhibition of channel activity was reversible since essentially full recovery of activity occurred with reoxygenation. Because the low pO_2 inhibited the potassium channel in the absence of cytosolic components, the oxygen sensing mechanism must be limited to the channel itself and/or the surrounding membrane. The results suggested that oxygen sensing involved non-heme iron containing proteins associated with channel molecules. Initially during hypoxia as ATP levels fall these channels would be activated thus hyperpolarizing the membrane. If oxygen levels decrease sufficiently the channels would be inhibited thus reducing potassium efflux causing membrane depolarization. How these two influences (low ATP and low oxygen levels) play out in the course of a hypoxic insult is unclear. ATP sensitive potassium channels (K_{ATP}) have been found in a variety of tissues and in general couple cell metabolic activity to the electrical activity of the cell (Ashcroft and Gribble, 1998). They are inhibited by physiological levels of ATP and they are also blocked by the sulphonylurea drugs. The K_{ATP} channel consists of two types of subunits: an inwardly rectifying K^+-channel subunit and a sulphonylurea receptor subunit. Both subunits are necessary to form a functional K_{ATP} channel and coassemble in an obligate 4:4 stoichiometry to form an octameric channel. The various K^+-channel and sulphonylurea subunits mix and match to form K_{ATP} channels with different pharmacological and nucleotide sensitivities (Ashcroft and Gribble, 1998). The role of ATP-sensitive potassium channels during hypoxia is even more complicated in light of the observation that neurons often respond to oxygen deprivation with an initial hyperpolarization followed by a depolarization when the anoxic insult is more prolonged (Haddad and Jiang, 1993a). This hyperpolarization is mediated by the activation of ATP-sensitive potassium channels and activation of these channels is thought to have a protective role in brain anoxia (Ballanyi, 2004; Haddad and Jiang, 1993a). How these channels are related to the ATP sensitive potassium channels described by Jiang and Haddad (1994) which are inhibited by hypoxia is unclear.

In fact, the whole area of potassium channels and their role in the depolarization induced by anoxia is quite complex. The channels described by Jiang and Haddad (1994) are ATP sensitive, voltage-activated as well as being calcium dependent. These channels are inhibited by hypoxia and such inhibition would result in membrane depolarization. Other ATP sensitive potassium channels are activated by hypoxia (probably due to the fall in ATP levels during hypoxia) and this activation leads to a hyperpolarization of the cell membrane (Ballanyi *et al.*, 1996; Haddad and Jiang, 1993a). To add to the complexity, Plant *et al.* (2002) have described potassium leak channels in cerebellar granule neurons that are inhibited by anoxia but unlike the channels described by Jiang and Haddad (1994) these channels lack voltage dependence. These particular potassium leak channels are inhibited by an acid pH and are members of the tandem pore, acid sensitive potassium channel (TASK) family. Plant *et al.* (2002) showed that inhibition of these channels by hypoxia resulted in a reversible depolarization of about 14 mV in cultured cerebellar granule cells. Hence, it is quite likely that hypoxia induced inhibition of potassium channels does contribute to the anoxic depolarization.

Most of the studies reviewed to this point have examined the effects of hypoxia/anoxia on ion gradients and membrane depolarization. Oxygen deprivation inhibits oxidative phosphorylation with a significant reduction in ATP production. However ATP production can still occur to a limited extent through anaerobic glycolysis. Ischemia can be considered to be a state of oxygen deprivation combined with glucose deprivation. During ischemia, aerobic and anaerobic metabolism is inhibited and ATP production virtually ceases. Loss of ion gradients with membrane depolarization occurs earlier in the course of ischemia than with oxygen deprivation alone (Hansen, 1985; Tanaka et al., 1997). A number of investigators have used an in vitro ischemia model (combined oxygen and glucose deprivation) to investigate the mechanism(s) underlying the loss of ion gradients and resultant membrane depolarization. Rat hippocampal CA1 neurons exposed to oxygen and glucose deprivation responded with an initial hyperpolarization followed by a slow depolarization which led to a rapid massive depolarization after about six minutes of ischemia (Tanaka et al., 1997). When the rapid depolarization occurred, the condition of the neurons became irreversible. Reintroducing oxygen/glucose did not reverse the rapid depolarization and the neurons did not show any functional recovery (Tanaka et al., 1997 and 1999). In fact when oxygen/glucose was reintroduced immediately after generation of the rapid depolarization, the neuron did not repolarize and the membrane potential became zero after about five minutes (Tanaka et al., 1997). In contrast, Sick et al. (1982b, discussed earlier) reported that rat parietal cortical neurons in vivo exposed to anoxia underwent a massive loss of intracellular potassium and no doubt a depolarization of cell membranes but that potassium ion homeostasis was restored when normal oxygen conditions were promptly reinstituted. In addition, although the rat brain became electrically silent within two minutes of anoxia, electrical activity fully recovered with resumption of oxygenation. Hence, loss of ion gradients with resultant membrane depolarization may be reversible when induced by oxygen deprivation alone, but such does not appear to be the case when the insult involves both oxygen and glucose deprivation.

Jarvis et al. (2001) showed that rat neocortical brain slices exposed to oxygen and glucose deprivation underwent an anoxic depolarization within about four to five minutes and that these neocortical neurons (compared to control tissue) displayed histological evidence of nuclear and cellular swelling. The anoxic depolarization could not be blocked by pre-treating neocortical slices with glutamate receptor (NMDA and non-NMDA) antagonists or with tetrodotoxin. Inhibition of the Na^+-K^+-ATPase pump did result in a response identical to the anoxic depolarization observed with oxygen/glucose deprivation.

Tanaka *et al.* (1997) examined the anoxic depolarization in adult rat hippocampal CA1 neurons exposed to oxygen/glucose deprivation. They made the following observations. The rapid depolarization was generated at the neuronal soma membrane as well as the membrane of apical and basal dendrites. This rapid depolarization was not reversed by reintroduction of oxygen/glucose even if this reintroduction occurred immediately. The free intracellular calcium concentration markedly increased during generation of the rapid depolarization and this rapid rise in intracellular calcium concentration correlated with the rapid depolarization. Blockade of glutamate receptors did not prevent the depolarization and in fact significant accumulation of extracellular glutamate only occurred after the anoxic depolarization. However, pretreatment with glutamate receptor blockers did allow the membrane potential to completely recover after reintroduction of oxygen and glucose. (Haddad and Jiang (1993a), as previously noted, also reported that glutamate receptors were not involved in the genesis of the anoxia induced depolarization.) The rapid depolarization was sodium, calcium and chloride dependent. The Na^+-K^+-ATPase pump inhibitor ouabain produced a rapid depolarization similar to that observed with oxygen/glucose deprivation. The conductances underlying the rapid depolarization could be activated whether the membrane was depolarized or hyperpolarized, i.e. its activation is voltage independent. They concluded that the rapid depolarization was voltage independent and probably due to a non-selective increase in the permeability of all participating ions. Tanaka *et al.* (1999) examined in more detail the calcium dependence of the ischemia induced rapid depolarization. They observed that if hippocampal CA1 neurons which had undergone a rapid depolarization were superfused with a calcium free medium (with inclusion of an inorganic calcium channel blocker) containing oxygen and glucose, the depolarization could be completely reversed. This reversal occurred only if the zero calcium, oxygen/glucose containing medium was superfused within one minute of the rapid depolarization. Neurons showed only partial or no recovery of the membrane potential if the zero calcium, oxygen/glucose containing medium was superfused after one and a half minutes. They also reported that hippocampal CA1 neurons underwent morphological changes following the rapid depolarization which included swelling of cell bodies and the appearance of small and later large blebs on the surface membrane of the cell body. In fact, small blebs were visible immediately after the rapid depolarization. These morphological changes were not altered by reintroduction of oxygen and glucose alone. However, if the neurons were superfused with calcium free (with inclusion of an inorganic calcium channel blocker) medium containing oxygen and glucose following the rapid depolarization, the morphological changes were partially prevented. In particular, these neurons did not demonstrate any blebs on the cell body surface membrane. These results suggest that calcium influx into the neuron immediately after generation of the rapid ischemic depolarization triggers irreversible membrane dysfunction associated with morphological evidence of membrane damage. The observation that blockade of NMDA and non-NMDA receptors allowed complete reversal of the depolarization with reintroduction of glucose/oxygen in the presence of a normal extracellular calcium concentration (Tanaka *et al.*, 1997) suggests that the ischemic depolarization activates glutamate receptors and that the increased influx of calcium occurs through ionotropic glutamate receptor channels. This suggestion is supported by the data of Bickler *et al.* (2000). These investigators showed that ten minutes of anoxia (replacement of oxygen by nitrogen in the perfusate) killed more than 85% of rat hippocampal CA1 neurons *in vitro*, but that the NMDAR (NMDA receptor) antagonist MK-801 prevented death of rat neurons. Tanaka *et al.* (1999) postulated that micro-pores formed in the small blebs and that these micro-pores were responsible for the rapid depolarization. A similar proposal has been advanced by Anderson *et al.* (2005) who suggested that the rapid depolarization was due to the opening of "megachannels." In fact, recently, Thompson *et al.* (2006) showed using rat hippocampal neurons that oxygen/glucose deprivation within about ten minutes opened large hemichannels formed from unapposed half gap junctions. Openings of these large conductance channels were rare under control conditions but the open probability was increased by a factor of 4.5 over control by oxygen/glucose deprivation. Thompson *et al.* (2006) proposed that ischemia opens large non selective neuronal gap junction hemichannels and that this is a central component of the increased membrane permeability underlying the dissipation of ion gradients during anoxic depolarization. However, since the report of Thompson *et al.* (2006), Douglas *et al.* (2006) have shown that pretreating neocortical rat brain slices with the gap junction/hemichannel blocker carbenoxolone did not prevent the anoxic depolarization induced by oxygen/glucose deprivation. Hence, while hemichannel opening clearly contributes to the large conductance increase observed following anoxic depolarization initiation, it does not appear to generate the anoxic depolarization itself.

The initial critical events occurring during anoxia or ischemia are dissipation of ion gradients resulting in a rapid depolarization of the cell membrane. Although the loss of ion gradients is clearly the initial event, for simplicity this whole process will be referred to as anoxic depolarization since this is the term frequently used in the literature.

The previous paragraphs have described a number of studies designed to examine the mechanisms underlying this anoxic depolarization. The following is a summary of the factors that have been observed to influence this depolarization process. This list is in no particular order: (1) The depolarization process is energy-linked and dependent upon the level of ATP stores. The onset of depolarization is delayed when adequate ATP levels can be maintained for a longer period of time. (2) The link to ATP levels involves functioning of the Na^+-K^+-ATPase pump since inhibition of this pump (under normoxic conditions) results in a rapid depolarization similar to that of anoxic depolarization. (3) Complete failure of the Na^+-K^+-ATPase pump however, cannot account for the rate of change in extracellular ion concentrations occurring during an anoxic or ischemic insult. A change in membrane ion permeability must also occur. (4) Blockade of glutamate receptors or application of the sodium channel blocker tetrodotoxin does not prevent anoxic depolarization. (5) Low tissue pO_2 levels can directly affect the activity of ion channels. For example hypoxia can open sodium channels but these particular channels are blocked by tetrodotoxin. Hence, their role in the genesis of anoxic depolarization is unclear. Hypoxia can also inhibit (close) potassium channels which results in membrane depolarization. (6) Anoxic depolarization appears to be a calcium dependent process. The rapid depolarization induced by oxygen/glucose deprivation cannot be reversed by reintroduction of oxygen and glucose. However the rapid depolarization can be completely reversed if oxygen and glucose are reintroduced to neurons superfused with a calcium free medium. (7) Large conductance neuronal gap junction hemichannels are opened by an ischemic insult but the factors linking the ischemic insult and opening of these channels are unknown. In summary, it is probably reasonable to conclude that currently there is no agreed upon explanation for the loss of ion gradients and resulting membrane depolarization. There does appear to be agreement that anoxia or ischemia opens up large conductance nonselective channels which allow ions to equilibrate across the membrane thus dissipating ion gradients and depolarizing the membrane. The signals that link the anoxic or ischemic insult to the opening of these channels are unknown. Perhaps one of the signals is falling ATP levels. In addition, there is still no agreement as to the actual nature of these anoxia/ischemia activated channels. In addition, it appears that the opening of these large channels with depolarization of the membrane also results in a significant influx of calcium into the cell. This increase in cellular calcium if large enough, appears to trigger irreversible membrane dysfunction.

With the onset of anoxia or ischemia, ATP production will decrease but ATP utilization will continue. ATP production will obviously be more severely impaired by oxygen/glucose deprivation than oxygen deprivation alone. With reduction in ATP levels the function of the Na^+-K^+-ATPase pump will become impaired and this pump will be less able to maintain ion gradients (specifically sodium and potassium gradients since this pump transports two potassium ions into the cell for every three sodium ions that are removed). In addition as noted in the "Brain Energy Metabolism," section, the Na^+-K^+-ATPase pump alone accounts for at least 50% of ATP turnover in mammalian cells. With prolongation of the anoxic/ischemic insult a rapid large increase in ion permeability occurs. As already discussed, what triggers this permeability increase and what are the signals linking the anoxic/ischemic insult and falling ATP levels to the opening of ion channels are unknown. The nature of the ion channels which are opened is unclear although they are probably of large conductance and non-selective. Opening of these channels would allow ions to equilibrate across the cellular membrane. Ion gradients would be dissipated and the membrane would be rapidly depolarized. In addition, the contribution that hypoxia induced activation of sodium channels and inhibition of potassium channels makes to the process of depolarization is unclear. The rapid anoxic depolarization is also associated with a rapid increase in intracellular calcium. The magnitude of this increase is greater with ischemia than with oxygen deprivation alone (Bickler and Hansen, 1994; Kulik et al., 2000). The sequence of events that results in the rapid anoxic depolarization is the critical process. In the case of ischemia, anoxic depolarization represents an irreversible stage and even immediate reintroduction of oxygen and glucose does not lead to functional recovery. This irreversibility appears to be triggered by the rapid increase in the intracellular calcium concentration that accompanies the rapid depolarization. When the insult is oxygen deprivation only, the depolarization can be reversed if normal oxygen conditions are quickly restored. At the time that anoxic depolarization occurs, ATP levels would be extremely low (more so with ischemia than with oxygen deprivation alone). The Na^+-K^+-ATPase pump would essentially be non-functional and hence unable to restore the ion gradients.

Loss of ion gradients with membrane depolarization triggers a cascade of effects which lead to cell injury and death. For example, the influx of sodium will be accompanied by the influx of chloride and water causing cell swelling and damage. Loss of the sodium gradient can reverse the usual exchange properties of the sodium/calcium exchanger resulting in calcium entering the cell. Membrane depolarization can activate voltage dependent calcium channels leading to increased calcium influx. With respect to the presynaptic membrane, depolarization can lead to a calcium dependent fusion of synaptic vesicles with the plasma membrane and release of neurotransmitters such as glutamate. An additional effect of the loss of the sodium gradient is reversal of sodium gradient dependent neurotransmitter re-uptake transporters thus further increasing the extracellular concentration of excitatory neurotransmitters such as glutamate (Bickler and Buck, 1998). This mechanism (reversal of neurotransmitter re-uptake transporters) is believed to be more important in increasing the extracellular concentration of glutamate than vesicular release. Vesicular release of glutamate fails early on in hypoxia as a result of the loss of ATP (Bickler and Buck, 1998). Activation of NMDA receptors by glutamate will lead to calcium inflow through the ion channels of these receptors. Hypoxia also induces calcium release from intracellular stores including mitochondria (Haddad and Jiang, 1993a). In fact, the elevation of intracellular calcium occurring during anoxia results in a number of deleterious intracellular events and intracellular overload of this cation is thought to be a key player in the genesis of cell injury (Haddad and Jiang, 1993a). This topic will be discussed in a subsequent section.

In summary, the anoxia/ischemia induced loss of ion gradients and resultant membrane depolarization represents a critical event which if not reversed results in a number of inter-related cellular processes (the two most important ones are believed to be the release of excitatory neurotransmitters and intracellular calcium overload) leading to cell death. However, the primacy of anoxic depolarization cannot be overemphasized. Despite this primacy, the actual mechanism(s) underlying the anoxic depolarization have not been established. The overview given in this section clearly does not due justice to the many factors that are believed to be involved in the genesis of cell injury and death during anoxia or ischemia. The topic of ischemic cell death has been extensively reviewed by Lipton (1999).

Anoxia tolerant mammalian neurons

Not all mammalian neurons are equally sensitive to anoxia and in this section the characteristics of anoxia tolerant neuron populations are reviewed.

Neonatal brain

It has been well documented that the neonatal brain is much more resistant to oxygen deprivation than that of the adult (Haddad and Jiang, 1993a). Hansen (1977) measured the extracellular potassium concentration in the parietal cortex of rats of varied ages (four days to adulthood) made anoxic by replacing inspired oxygen with nitrogen. Under normoxic conditions, the extracellular potassium concentration was similar in all age groups and varied from 2.9 to 4.6 mM. Extracellular potassium concentration rose following nitrogen inhalation and the pattern was similar in all age groups. There was an initial slow rise in potassium concentration which was followed by a sudden steep rise to about 70 mM. After that the concentration increased more slowly to a final value of about 90 mM. The significant difference between the age groups was the time until the steep rise occurred. In the four day-old rat the mean time was about 20 minutes whereas in the adult the time was about one and a half minutes. Rats that were seven, 12, 16 or 24 days old had times that fell progressively between the 20 and one and a half minute extremes for the four day-old and adult rat. In adult rats that were reoxygenated just after the steep increase had occurred, the extracellular potassium concentration returned to preanoxic levels in most cases.

Haddad and Donnelly (1990) reported that in hypoglossal motoneurons, the membrane depolarization induced by hypoxia varied with age. Brain slices were prepared from rats of various ages and anoxia induced by replacing oxygen with nitrogen in the perfusate. Rats three to seven days old showed a change in membrane potential which was about one third of that observed in the adult (about 10 mV versus 30 mV). In addition, as adult neurons depolarized they became spontaneously active whereas none of the neonatal neurons became spontaneously active. Thus neonatal hypoglossal neurons showed a smaller depolarization and very little spontaneous neuronal activity when exposed to hypoxia compared to adult hypoglossal neurons. Thus less depolarization and less spontaneous neuronal activity as observed in the neonatal neuron compared to the adult neuron would preserve intracellular ATP and lead to a much longer maintenance of ionic homeostasis (Xia et al., 1992).

Bickler et al. (1993) examined the effects of anoxia on intracellular (free) calcium regulation in cerebral cortex brain slices from rats of various ages. Chemical anoxia induced with cyanide induced only modest increases in intracellular calcium in rats less than 15 days old but significant increases in intracellular calcium in older rats. There was in fact, a direct linear relationship between peak intracellular calcium concentration induced by cyanide and postnatal age of the rat. However, neonatal rat brain slices treated with both cyanide and iodoacetate (an inhibitor of glyceraldehyde-3-phophate dehydogenase) to block both aerobic and anaerobic metabolism displayed increases in intracellular calcium equivalent to that of the adult. Application of glutamate under normoxic conditions caused a greater increase in intracellular calcium in brain slices from the adult compared to brain slices from the neonate. In fact, maximum elevation in brain slice intracellular calcium produced by glutamate increased significantly with postnatal age. The glutamate receptor antagonist MK-801 significantly reduced the magnitude of the cyanide induced increase in intracellular calcium in brain slices from rats 14 days old or younger but had no significant effect on the cyanide induced increase in intracellular calcium in slices from adult rats. Cyanide exposure reduced ATP levels by 50% in neonatal rat brain slices after ten minutes, but caused a greater than 90% reduction after ten minutes in brain slices from rats older than 14 days. In slices from rats older than 20 days ATP levels approached zero after ten minutes of cyanide exposure. These differences in ATP levels induced by cyanide in rats of different ages probably reflect differences in ATP utilization since neonatal rat brain slices showed a more gradual decline in ATP levels following blocking of both aerobic and anaerobic metabolism than adult rat brain slices. In a more recent paper, Bickler and Hansen (1998) studied the properties of hippocampal CA1 neurons in brain slices from rats ages 3–30 postnatal days (PND). Hypoxia was induced by replacing oxygen in the bathing medium with nitrogen. Hypoxia induced cell death was assessed in slices that had undergone a five-minute exposure to hypoxia followed by six hours of recovery in oxygenated medium. The viability and morphology of neurons from rats 3–7 PND was not significantly affected by five minutes of anoxia, whereas the same period of anoxia in rat neurons 18–22 PND resulted in significant morphological cell injury and cell death. Morphological changes included bleb formation, cell swelling and disintegration. Intracellular free calcium increased more significantly with hypoxia in 18–22 PND rat neurons than in 3–7 PND rat neurons. In fact, there was a direct linear relationship between maximal change in intracellular calcium concentration induced by hypoxia and postnatal age of the rat. In contrast to the study by Bickler *et al.* (1993) blocking both aerobic and anaerobic metabolism (with hypoxia, cyanide and iodoacetate) caused a greater increase in intracellular free calcium in 18–22 PND rat neurons, but had little effect on 3–7 PND neurons. Glutamate release from hippocampal slices during hypoxia, cyanide poisoning or depolarization with potassium chloride increased significantly with increasing postnatal age. Hypoxia induced release of glutamate from both 3–7 PND and 18–22 PND slices was not due to synaptic vesicular release. A combination of NMDA, AMPA and metabotropic glutamate receptor antagonists reduced the hypoxia induced increase in intracellular free calcium by 62% in slices from 18–22 PND rats but only 33% in slices from 3–7 PND rats. This observation is also different from the result reported by Bickler *et al.* (1993). Those investigators observed that blocking of NMDA receptors significantly reduced the cyanide induced increase in intracellular calcium in neonatal rat brain slices but had no effect in brain slices from rats older than two weeks. In the neonatal brain, ATP utilization and depletion rates during hypoxia are reduced and greater amounts of ATP would remain available for ion homeostasis as pointed out by Duffy *et al.* (1975) and Bickler *et al.* (1993). Hence in the neonate, better maintenance of ion gradients will favor continued functioning of sodium gradient dependent glutamate re-uptake transporters and less glutamate will accumulate in the extracellular space. According to Bickler and Hansen (1998) this reduced glutamate concentration coupled with a previously demonstrated lower activity (or possibly fewer numbers) of glutamate receptors will mean that the neonate will show less glutamate mediated calcium influx during hypoxia than the adult. Although the studies by Bickler *et al.* (1993) and Bickler and Hansen (1998) generally gave comparable results there were differences as noted. These differences may reflect the fact that these two sets of investigators used different regions of the brain for their studies.

In a more recent study, Bickler *et al.* (2003) examined the oxygen sensitivity of NMDA receptors as a function of age of the animal (rat). Rats of between 2–30 days old (P2–30) were used in these experiments. Both hippocampal brain slices and dissociated CA1 neurons were studied. Hypoxia was induced by using gas mixtures with differing mixtures of nitrogen and oxygen. The functional activity of NMDA receptors (NMDAR) was assayed by measuring the increase in free intracellular calcium induced by the receptor agonist NMDA under conditions of normal oxygenation and hypoxia. Hypoxia (pO_2 10 mmHg) reversibly decreased NMDAR activity in P3–10 CA1 neurons by a mean reduction of 68% (i.e. the NMDA induced increase in intracellular calcium during hypoxia was only 32% of the value obtained during normal oxygenation) compared to a smaller depression (mean value of 20%) in older, P18–22, neurons. Consistent with this result, the investigators using cell attached patch clamping, observed that hypoxia reduced the open probability of NMDAR in P4–9 neurons by 38% (mean value) whereas no reduction in open probability of NMDAR was obtained in P21 neurons. Hypoxia also protected neonatal neurons from NMDA induced toxicity. Application of NMDA to P3–6 neurons under normal oxygenation resulted in the death of 65% of neurons. Application of NMDA to neurons of similar age during hypoxia resulted in the death of only 35% of cells. In neurons of age P17–21, application of NMDA resulted in the death of about 32% of neurons. During hypoxia, NMDA induced a statistically similar percentage of cell death (24%) in P17–21 neurons. Hence, although hypoxia protected neonatal but not mature neurons from NMDA toxicity, NMDA induced more toxicity in neonatal neurons than in mature neurons. The explanation for this finding lies in the experimental observation of Bickler *et al.* (2003) that there is an inverse relationship between the magnitude of the NMDA stimulated increase in intracellular calcium (under normal oxygenation) and postnatal age. Finally, the investigators found at least two reasons for the hypoxia induced decrease in NMDAR activity in neonatal neurons compared to older neurons. In some neurons hypoxia induced a modest increase in intracellular calcium which was often associated subsequently with a reduced response to NMDA. By using an actin microfilament stabilizer, these investigators demonstrated that the increase in intracellular calcium induced by hypoxia led to reduced NMDAR activity as a result of a calcium dependent alteration in actin polymerization. A second explanation involved the difference in subunit composition of NMDAR in neonatal and more mature CA1 neurons. NMDAR in neonatal neurons contain the subunits NR1/NR2D or NR1/NR2B, and NMDR in mature neurons contain the subunits NR1/NR2A or NR1/NR2C. In oocytes expressing NR1 plus NR2D subunits, NMDA elicited currents were reduced by hypoxia whereas oocytes expressing NR1/NR2C receptors showed an augmentation of NMDA elicited currents during hypoxia. These observations suggest direct effects of oxygen on the receptor proteins and indicate that in the neonatal brain hypoxia induces a reduction in membrane ion permeability. Such a reduction in ion permeability would result in a decrease in ATP utilization since less active ion pumping would be necessary to maintain ion gradients.

The metabolic differences between fetal and postnatal rat brain in relation to anoxia survival have been studied by Duffy *et al.* (1975). Anoxia was induced by placing rats in jars containing a nitrogen atmosphere. The effects of ischemia were examined as follows. Rats were decapitated and the severed heads were maintained at 37°C and frozen in liquid nitrogen at specific time periods after the decapitation. Forebrains were removed and the levels of various substrates measured. The survival times of rats in a nitrogen atmosphere were inversely proportional to age. The LD85 was 45 minutes for term fetal rats, 25 minutes for one-day-old rats, 8.8 minutes for seven-day-old rats and only 0.8 to 1 minute in adult rats. After ten minutes of ischemia, forebrain ATP levels had fallen by 32% in the one-day-old rat, 64% in the fetus and 90% in the seven-day-old rat. From the measurement of levels of phosphocreatine, ATP, ADP, glucose and lactate at time zero and specific time points after decapitation (up to ten minutes), the investigators were able to calculate energy utilization rates during ischemia. The rates (in mmol high energy P/kg brain/min) were: term fetus 1.57, one-day-old 1.33, seven-day-old 2.58, and for comparison the value in the adult rat brain is 26.8. Hence, a major factor explaining the longer survival during anoxia of fetal and one-day-old rats compared to seven-day-old and adult rats is their much lower rates of cerebral energy utilization during anoxia.

Xia et al. (1992) compared oxidative and glycolytic capacity in the brains of newborn (5-day old) and adult rats. These investigators measured cytochrome c oxidase activity as an index of oxidative capacity and hexokinase activity as an index of glycolytic capacity. Cytochrome c oxidase activity was higher in the newborn but hexokinase activity was comparable in the newborn and adult. These results would be consistent with the observation of Duffy et al. (1975) that the neonate reduces energy utilization during anoxia rather than increasing anaerobic ATP production. As suggested by the experimental observations of Haddad and Donnelly (1990) already noted, the reduced energy utilization of neonatal neurons during hypoxia (compared to adult neurons) may be due to a reduction in spontaneous electrical activity during hypoxia in neonates such that less energy is consumed maintaining ion homeostasis.

Dorsal vagal neurons

Dorsal vagal neurons (DVN), which are involved in the nervous control of autonomic functions, have a similar high tolerance to oxygen depletion as do the neurons of neonates (Ballanyi and Kulik, 1998). Exposure of DVN to five minutes of hypoxia (replacing oxygen in the bathing medium with nitrogen) led to a very modest increase in extracellular potassium concentration (2.3 to 3.4 mM) and a very small decrease in extracellular pH (Ballayni et al., 1996). Similarly, DVN exposed to glucose free medium showed only minor increases in extracellular potassium concentration and an increase in extracellular pH. However, DVN exposed to both hypoxia and glucose deprivation displayed a larger change in extracellular potassium with an increase in concentration to 11 mM. However, by way of comparison, transient rises in extracellular potassium concentration up to 9 mM can be observed when DVN are repetitively stimulated. In glial cells of the dorsal vagal nucleus, hypoxia or glucose deprivation led to a mean depolarization of 9 mV whereas hypoxia combined with glucose deprivation resulted in a mean depolarization of about 40 mV. In 38% of DVN, hypoxia of five to 20 minutes duration produced a stable hyperpolarization of about 15 mV while in 25% of DVN similar periods of hypoxia evoked a stable depolarization of less than 10 mV. In 37% of DVN, anoxic exposure did not lead to any significant change in membrane potential. DVNs that hyperpolarized during anoxia showed a very similar hyperpolarization when exposed to glucose free medium. This observation of anoxia induced hyperpolarization is at odds with the report of Haddad and Jiang (1993a) who stated that no hyperpolarization had been observed in studies of more than 350 to 400 brainstem neurons in their laboratory. In addition, according to Haddad and Jiang (1993a) hippocampal and neocortical neurons do show an initial hyperpolarization with anoxia but this hyperpolarization is not stable and is followed by a depolarization. By contrast, in DVN the hyperpolarization induced by anoxia or glucose deprivation is quite stable and not followed by depolarization (Ballayni *et al.*, 1996). These oxygen or glucose deprivation induced hyperpolarizations appeared to be mediated by ATP-sensitive potassium channels (K_{ATP}) since blockade of K_{ATP} channels with tolbutamide completely reversed the (hyperpolarization) change in membrane potential. Ballanyi *et al.* (1996) also examined the combined effects of hypoxia and glucose deprivation. In the DVN studied, hypoxia or glucose deprivation alone resulted in a hyperpolarization. However when hypoxia and glucose deprivation were combined a rapid progressive depolarization occurred accompanied by a decrease in membrane resistance. This depolarization was completely reversed by reintroduction of glucose and oxygen (how soon this was done is not stated) which differs from the situation in hippocampal neurons *in vitro*. In these neurons the ischemia induced rapid depolarization was not reversed by reintroduction of glucose and oxygen and these neurons showed no functional recovery (Tanaka *et al.*, 1997 and 1999). Hence neurons in the dorsal vagal nucleus are relatively resistant to anoxia or glucose deprivation and do not respond to either of these insults with a progressive depolarization or with significant increases in extracellular potassium concentration. However, ischemia (anoxia plus glucose deprivation) does lead to a progressive depolarization and a more significant disturbance in ion homeostasis although these changes are reversed by reintroduction of oxygen/glucose.

In a more recent study, Kulik *et al.* (2000) examined the effects of anoxia and ischemia on calcium homeostasis in dorsal vagal neurons (DVN). Anoxia induced membrane hyperpolarization and a concomitant small increase in intracellular calcium. To exclude the effects of voltage activated calcium channels, the anoxic response was studied under voltage clamp conditions. A rise in free intracellular calcium still occurred and was accompanied by a sustained outward current. Tolbutamide blocked the hyperpolarization and outward current but not the rise in intracellular calcium. Tetrodotoxin and ionotropic glutamate receptor antagonists also failed to block the anoxia induced rise in intracellular calcium. The outward current and elevation of intracellular calcium were unaffected by removing calcium from the bathing medium. Hence the anoxia induced increase in intracellular calcium is probably due to calcium release from intracellular stores. Ischemia (glucose and oxygen deprivation) evoked a progressive depolarization (by about 30 mV) accompanied by a much larger increase in intracellular calcium than induced by anoxia. The membrane depolarization and rise in intracellular calcium were reversed by reoxygenation even without glucose replacement. Again to exclude the effects of voltage activated calcium channels voltage clamp conditions were used. The rapid rise in intracellular calcium still occurred and was accompanied by a strong inward current. Recovery from the ischemic rise in intracellular calcium depended upon immediate reoxygenation after onset of the (ischemic) inward current. Both the rapid rise in intracellular calcium and the large inward current were significantly reduced by ionotropic glutamate receptor antagonists. The investigators concluded that ischemia resulted in an increase in extracellular glutamate initially due to calcium dependent exocytosis. A reversed action of glutamate receptors did not appear to play a major role in the early phase of ischemic glutamate release. The glutamate evoked increase in intracellular calcium was caused by calcium influx through ionotropic glutamate receptors. Calcium influx also occurred through voltage activated calcium channels. The large inward (ischemic) current probably represents inward flow of calcium. In summary, this study showed that DVN exposed to anoxia alone demonstrated only a small increase in intracellular calcium that was not mediated by glutamate receptors. Presumably extracellular glutamate concentrations were not increased during oxygen deprivation alone. On the other hand, ischemia resulted in a much greater increase in intracellular calcium which was mediated to a large extent by glutamate receptors implying that ischemia did result in accumulation of extracellular glutamate. The experimental data suggested that the increase in extracellular glutamate was due, at least initially, to calcium dependent exocytosis.

In summary, what are the adaptations characteristic of anoxia tolerant mammalian neurons?

Neonatal neurons

Compared to mature neurons, neonatal neurons can tolerate longer periods of anoxia before anoxic depolarization occurs and can maintain higher levels of ATP during hypoxia or ischemia due to a reduction in energy utilization. Higher levels of ATP in the neonatal neuron will allow continued functioning of the Na^+-K^+-ATPase pump which will help maintain ion gradients (specifically sodium and potassium gradients). When anoxic depolarization occurs, neonatal neurons demonstrate a much smaller change in membrane potential than adult neurons. However the mechanisms underlying these differences between neonatal and mature neurons are unknown. As discussed in the "General Features of Neuronal Response" section, there is currently no agreed upon model accounting for the rapid loss of ion gradients with depolarization of the membrane (referred to as anoxic depolarization) that occurs as a result of hypoxia or ischemia. Although maintenance of ATP levels allows neonatal neurons to tolerate longer periods of hypoxia/ischemia than mature neurons, the mechanism linking ATP levels and hypoxia tolerance is not well understood. Continued functioning of the Na^+-K^+-ATPase pump is clearly important but in itself is not a sufficient explanation for the ability of neonatal neurons to tolerate longer periods of hypoxia before undergoing anoxic depolarization. When neonatal neurons do undergo anoxic depolarization why is the change in membrane potential less than that observed in mature neurons? The answer is unknown but clearly must be related to differences between neonatal and mature neurons with respect to the nature of the large channels responsible for anoxic depolarization and/or the factors opening these channels during hypoxia/ischemia. Neonatal and mature neurons also differ with respect to the effects of hypoxia on calcium homeostasis. Hypoxia induces an increase in the intracellular concentration of calcium but the magnitude of this increase is much greater in mature neurons than in neonatal neurons. However, experiments designed to examine the mechanism of this difference have given conflicting results. Some examples follow. How is glutamate receptor activity related to neuronal "age"? Bickler *et al.* (1993) showed that the peak intracellular calcium concentration induced by the application of glutamate (under normoxic conditions) increased with rat postnatal age suggesting that glutamate receptor activity was greater in mature neurons than in neonatal neurons. On the other hand, Bickler *et al.* (2003) reported that the NMDA stimulated peak increase in intracellular calcium concentration (under normoxic conditions) decreased with rat postnatal age implying that glutamate receptor activity was less in mature than neonatal neurons. What is the effect of NMDA receptor antagonists? Bickler *et al.* (1993) reported that the NMDA receptor antagonist MK-801 was more effective in reducing the hypoxia (induced with cyanide) stimulated increase in intracellular calcium in neonatal than in mature neurons. On the other hand, Bickler and Hansen (1998) showed the opposite result. MK-801 was more effective in reducing the hypoxia (produced by oxygen deprivation) stimulated increase in intracellular calcium concentration in mature than in neonatal neurons. Do these opposite results reflect differences in experimental designs? Bickler *et al.* (1993) used cerebral cortical brain slices and induced hypoxia with cyanide. Bickler and Hansen (1998) used hippocampal brain slices and induced hypoxia by replacing oxygen with nitrogen in the bathing medium. In the studies of Bickler *et al.* (2003) hippocampal neurons were also used. However, it is uncertain how to reconcile these differing results and hence it is difficult to know what role glutamate receptors (especially NMDA receptors) play in explaining the differences in the effects of hypoxia on calcium homeostasis between neonatal and mature neurons.

In essence, the two major adaptations of neonatal neurons to anoxia (compared to adult neurons) are: reduced utilization of ATP during anoxia and a resistance to anoxia induced depolarization.

Dorsal vagal neurons (DVN)

These neurons appear to tolerate up to 20 minutes of hypoxia without undergoing a progressive depolarization. In about 40% of these neurons, hypoxia induces a stable hyperpolarization due to the opening of ATP sensitive potassium channels. However in the other 60% of neurons, hypoxia results in either a modest stable depolarization or no significant change in membrane potential. The reasons why these neurons do not demonstrate a progressive depolarization with hypoxia are unknown. As with neonatal neurons, one could speculate that the reasons relate to differences between DVN and other neuron populations in the nature of the channels responsible for the anoxic depolarization and/or in the factors opening these channels during hypoxia. In these neurons, hypoxia induces only a small increase in intracellular calcium which is due to release from intracellular stores and is not mediated by glutamate receptors. In other words, hypoxia does not appear to activate glutamate receptors in these neurons. When glucose and oxygen deprivation (ischemia) are combined, DVN undergo a progressive depolarization and also demonstrate a much more pronounced increase in the intracellular calcium concentration. The ischemic induced rise in intracellular calcium is due in part to activation of glutamate receptors with increased influx of calcium. However, in DVN neurons the ischemic depolarization and ischemic increase in intracellular calcium are reversible with prompt reintroduction of oxygen/glucose, whereas in hippocampal CA1 neurons the ischemic depolarization is not reversed with immediate reintroduction of oxygen/glucose (Tanaka *et al.*, 1997 and 1999).

Similar than to neonatal neurons, DVN neurons appear to be resistant to hypoxia induced depolarization. There are no know data as to rates of energy utilization during hypoxia in DVN neurons.

Intracellular calcium

As has been discussed, hypoxia and ischemia result in significant increases in the concentration of free intracellular calcium. During hypoxia, the extent of the rise in intracellular calcium varies with the neuron population being less in hypoxia tolerant neurons such as those from neonates and in dorsal vagal neurons. There are a number of mechanisms responsible for the hypoxia induced increase in intracellular calcium (Haddad and Jiang, 1993a). There can be increased influx of calcium through voltage gated calcium channels and through excitatory neurotransmitter (glutamate) activated channels. Calcium can also be released from intracellular stores including mitochondria, endoplasmic reticulum and nucleus. In addition declining ATP levels during hypoxia would result in inhibition of the ATPase calcium pump with impairment of calcium efflux from the cell. With dissipation of the sodium gradient, reversed functioning of the calcium/sodium exchanger can occur, leading to actual influx of calcium into the cell. Experimental studies have been performed to quantify the magnitude of these different mechanisms but the results have not been consistent.

Bickler and Hansen (1994) showed that in rat cortical (temporal-parietal) slices, ischemia (cyanide plus iodoacetate) produced a much greater increase in the free intracellular calcium concentration than hypoxia (oxygen deprivation or cyanide alone). In addition, the hypoxia or ischemia induced increase in free intracellular calcium was due predominantly to calcium influx since removing calcium from the bathing medium reduced the increase in intracellular calcium by over 85%. They also reported that antagonists of glutamate receptors, voltage gated sodium or calcium channels or calcium-sodium exchangers caused at best only modest reductions in the magnitude of the hypoxia or ischemia induced increase in intracellular calcium. They postulated that the bulk of the increased calcium influx occurred through areas of membrane damage. In a later paper, Bickler and Hansen (1998; see "Neonatal Brain" section in "Anoxia Tolerant Mammalian Neurons") reported that in hippocampal CA1 mature neurons a combination of NMDA, AMPA and metabotropic glutamate receptor antagonists reduced the hypoxia induced increase in free intracellular calcium by 62%. In dorsal vagal neurons hypoxia induces only a modest increase in free intracellular calcium which is thought to originate via release from intracellular stores. Ischemia in dorsal vagal neurons causes a much greater rise in free intracellular calcium which is due to increased calcium influx through NMDA receptor and voltage gated calcium channels (Kulik *et al.*, 2000). This small sampling of studies should suffice to point out that the relative importance of glutamate receptors, voltage gated calcium channels or calcium/sodium exchangers in mediating the hypoxia or ischemia induced increase in free intracellular calcium has not been established. One possible explanation for this lack of consistent results is that the same neuron population has not been used in these various studies. These different neuron populations (hippocampal CA1 neurons, cortical neurons, dorsal vagal neurons) appear to differ in terms of their hypoxia tolerance and may also differ with respect to the effects of hypoxia on their ability to regulate calcium homeostasis.

As reviewed by Haddad and Jiang (1993a), intracellular calcium overload has been implicated in the genesis of hypoxia/ischemia induced cell damage through the activation of a variety of intracellular enzymes. However, Haddad and Jiang (1993a) pointed out that anoxia induced cell damage in rat (hippocampal) CA1 cells occurred even if the increase in intracellular calcium was prevented by using a calcium free bathing medium. On the other hand, the studies of Tanaka *et al.* (1997 and 1999; see "Oxygen and/or Glucose Deprivation and the Brain" section in "General Features of Neuronal Response") do implicate increased free intracellular calcium in the development of irreversible membrane dysfunction. In hippocampal CA1 neurons, ischemia (oxygen/glucose deprivation) results in rapid massive depolarization which is irreversible even if oxygen and glucose are promptly reintroduced. In addition, this rapid depolarization is associated with morphological evidence of membrane damage. The rapid depolarization can be completely reversed by reintroduction of oxygen and glucose if the neurons are also bathed in a calcium free medium. Not only does the membrane potential recover under these conditions, there is also partial prevention of the morphological membrane changes. However, in the presence of a calcium free medium, the glucose and oxygen must be reintroduced within one minute following the rapid depolarization for reversal to occur. Hence somehow the rapid increase in intracellular calcium accompanying the ischemic induced rapid depolarization triggers irreversible membrane damage and dysfunction. Tanaka *et al.* (1999) were somewhat circumspect with respect to the mechanism of the depolarization induced increase in intracellular calcium. They suggested that the increase in intracellular calcium was due to enhanced influx through damaged membrane areas and through ionotropic glutamate receptor channels as well as due to increased release of calcium from internal stores.

In summary, the rapid membrane depolarization induced by hypoxia or ischemia is associated with a rapid increase in the free intracellular calcium concentration. Ischemia induces a greater increase in intracellular calcium than oxygen deprivation alone. Experimental observations indicate that this rise in intracellular calcium is a consequence and not the cause of the anoxic depolarization. Dissipation of the sodium gradient precedes the rise in intracellular calcium (Hammarstrom and Gage, 2002) and the anoxic depolarization is not prevented by bathing neurons with medium containing a low calcium concentration (Tanaka *et al.*, 1997). The rise in intracellular calcium appears to be predominantly due to increased influx of extracellular calcium with a modest contribution coming from the release of intracellular calcium stores. The relative importance of various calcium permeable channels (including glutamate receptor channels), reversed functioning of calcium-sodium exchangers (due to loss of the sodium gradient) and inhibition of ATPase calcium pumps in mediating the enhanced calcium influx has not been established. In the case of ischemia, the rapid influx of extracellular calcium following the anoxic depolarization appears to trigger irreversible membrane damage and dysfunction. Reintroduction of oxygen and glucose even immediately following the anoxic depolarization does not reverse the depolarization nor lead to neuronal recovery. In the case of hypoxia (oxygen deprivation), reversal of the anoxic depolarization is generally possible if oxygen is reintroduced promptly. This observation suggests that with hypoxia, unlike ischemia, the increase in intracellular calcium accompanying the anoxic depolarization does not immediately set in motion the sequence of events leading to irreversible membrane dysfunction. However, the time window of reversibility for hypoxia has not been established, i.e. how soon following the hypoxia induced rapid depolarization must reoxygenation occur before the depolarization becomes irreversible. In addition, this whole area of reversibility is complicated by the observation that different neuron populations have different levels of hypoxia tolerance thereby making generalizations difficult. As already noted, in dorsal vagal neurons, the rapid depolarization induced by ischemia is reversed by the prompt return of oxygen and glucose (Ballanyi *et al.*, 1996), whereas in hippocampal CA1 neurons the ischemic induced rapid depolarization is not reversed even when oxygen and glucose are immediately reintroduced (Tanaka *et al.*, 1997 and 1999).

Glia

Glial cells, non-excitable cells that possess extensive processes but lack axons and dendrites, far outnumber neurons in the brain and play crucial roles in many aspects of the central nervous system. Astrocytes detect changes in the activity of their adjacent neurons through ion channels, receptors and transporters expressed on their surface and in turn, respond by releasing a variety of substances known as gliotransmitters (Panatier *et al.*, 2006). One of these gliotransmitters is the amino acid D-serine and through release of this amino acid, astrocytes regulate the activation of synaptic NMDA receptors and therefore glutamatergic transmission at the postsynaptic level as follows. NMDA receptors require a co-agonist along with glutamate to open an intrinsic ion-permeable pore. This co-agonist can be glycine or D-serine (Diamond, 2006). D-serine is synthesized in astrocytes and released through an exocytotic pathway. Panatier *et al.* (2006) showed that astrocytes could control the concentration of D-serine in the synaptic cleft and consequently the extent to which D-serine sites on synaptic NMDA receptors were occupied; thus regulating the activity of these receptors. Glial-neuron interactions can also be mediated through gap junctions. Gap junctions are clusters of intercellular channels that when open can provide conduits for diffusional exchange of ions and small molecules directly from one cell to another. Gap junctions mediate communication between glial cells, but there is also evidence that gap junctions can mediate intercellular communication between astrocytes and neurons (Farahani *et al.*, 2005). Gap junctions may play a role in ischemia mediated neuronal cell death (in addition to the role played by unapposed half gap junction hemichannels previously discussed with respect to the process of anoxic depolarization). There is some evidence that gap junctions may facilitate the intercellular transfer of intracellular calcium and other "toxic" metabolites from ischemia damaged astrocytes to attached adjacent neurons (Farahani *et al.*, 2005). However, the evidence supporting a role for gap junction communication between astrocytes and neurons in the genesis of anoxia/ischemia brain damage is still very incomplete.

Another mechanism by which glial cells may participate in neuronal cell damage has been described by Yamashita *et al.* (2006). These investigators studied the effect of five minutes of global ischemia (cardiac arrest) on the survival of cerebellar Purkinje cells in mice lacking the glial glutamate-aspartate transporter (GLAST). After the five minutes of cardiac arrest, the mice were resuscitated and the histology of the cerebellum examined five days later. This transporter is localized predominantly to glial cells and is one of five subtypes whose function is to remove glutamate from the extracellular fluid in the brain. In wild type mice subjected to five minutes of ischemia, no definite Purkinje cell loss was found, whereas in mice lacking GLAST a significant (ischemia induced) loss of Purkinje cells was observed. The investigators concluded that GLAST was indispensable for preventing excitotoxic cerebellar damage after ischemia. Also, it should be noted that Yamashita *et al.* (2006) limited the period of cardiac arrest to five minutes, since longer periods resulted in such a high mortality rate in the mice that the study was technically not possible.

This brief discussion should illustrate the complexity of glial-neuronal interactions and their potential importance in hypoxia/ischemia neuronal cell damage. Hypoxia/ischemia also damages glial cells directly since oxygen deprivation and oxygen/glucose deprivation both lead to glial cell membrane depolarization (Ballanyi *et al.*, 1996) and Farahani *et al.* (2005) point out that astrocytes are probably as susceptible to hypoxic or ischemic cell damage as neurons.

Comparative effects of ammonia and hypoxia/ischemia on neurons

There are many parallels between the neurotoxic effects of ammonia and hypoxia/ischemia. In their extensive review of the biochemistry and physiology of brain ammonia, Cooper and Plum (1987) devoted a brief section to the comparative effects of excess ammonia and ischemia on cerebral energy metabolism. Since the topic of ammonia toxicity is covered in detail in Chapter 5, some of these similarities are just highlighted in this section.

Both excess ammonia and hypoxia/ischemia decrease brain ATP levels. In both cases, ATP levels fall as a result of decreased production with continued or increased utilization. The underlying mechanisms by which excess ammonia and hypoxia/ischemia deplete ATP levels do not appear to be the same. In the case of hypoxia/ischemia, ATP production is inhibited by a lack of substrates (oxygen and glucose) whereas in the case of ammonia toxicity, impaired ATP production is believed to be secondary to impairment of mitochondrial function by disordered calcium homeostasis. Activation of glutamate receptors, particularly NMDA receptors, is thought to play a pivotal role in mediating ammonia neurotoxicity. The current working model is that ammonia activates NMDA receptors by depolarizing the postsynaptic membrane with release of the magnesium block thus allowing increased activation of the receptors without an increase in extracellular glutamate concentration. Ammonium causes membrane depolarization directly by being able to partially substitute for either sodium or potassium (Binstock and Lecar, 1969). Activation of NMDA receptors allows increased entry of calcium and sodium into the cell. Increased intracellular calcium is probably an important mediator of the subsequent events believed to be responsible for the neurotoxic manifestations of excess ammonia. Hypoxia/ischemia also leads to membrane depolarization although the underlying mechanism is still not well understood. The anoxic depolarization is accompanied by a rapid influx of extracellular calcium and increased intracellular calcium is believed to be an important factor in the genesis of hypoxia/ischemia induced neuronal damage. However, the relative importance of glutamate receptors in mediating the hypoxia/ischemia induced increased influx of calcium has not been clarified. In summary, both ammonia excess and hypoxia/ischemia lead to depletion of ATP and an increase in intracellular calcium concentration. Glutamate receptors appear to play a much more important role in mediating the "toxic" effects of excess ammonia compared to hypoxia/ischemia. Finally, morphological damage is most apparent in the glial cell in the case of ammonia toxicity whereas both the neuron and glial cell are damaged by hypoxia/ischemia.

As discussed in Chapter 5, fish in general and certain species in particular, are very ammonia tolerant compared to mammals. Most likely the mechanisms underlying the neurotoxic manifestations of excess ammonia and hypoxia/ischemia share many more common features than has currently been established. With this speculation in mind, it would be very interesting to determine whether ammonia tolerant fish are also hypoxia tolerant. If these two features, ammonia tolerance and hypoxia tolerance, do indeed go together than this finding would support the notion that excess ammonia and hypoxia/ischemia probably cause neurotoxicity via a common mechanism.

Chronic mild hypoxia and the brain

In the previous sections, the effects of severe acute oxygen or oxygen/glucose deprivation on neuronal function have been reviewed. Within minutes after the onset of the hypoxic or ischemic insult, neurons undergo a rapid (anoxic) depolarization which if not reversed leads to cell death. In the case of "milder" states of hypoxia (arterial pO_2 above approximately 45 mmHg), the central nervous system (CNS) displays a number of adaptations if this state is prolonged. Residing at a high altitude would be the most obvious example of circumstances in which the CNS is exposed to a chronic mild hypoxic state. These adaptations include increased ventilation, a fall in core body temperature, right shift of hemoglobin dissociation curve, increased packed cell volume (hematocrit), and vascular remodeling (change in density of capillaries) within the brain (LaManna *et al.*, 2004).

LaManna *et al.* (1992) studied rats exposed to three weeks of hypobaric (0.5 atm) hypoxia. Placement of the rats in the hypobaric chamber resulted in a fall in the arterial pO_2 to about 40 mmHg within 15 minutes. At the end of three weeks, the hypoxic group had a mean arterial pO_2 of 53 mmHg, a mean arterial pCO_2 of 18 mmHg and a mean hematocrit of 71%. In the control group (three weeks of normobaric, normoxic conditions) the corresponding values were: mean arterial pO_2 98 mmHg, mean arterial pCO_2 33 mmHg, and mean hematocrit 48%. Cerebral blood flow (measured in five regions) was increased after 15 minutes of hypoxia and remained elevated after three hours of hypoxia, although not to the same extent as that measured at 15 minutes. This early increase in cerebral blood flow was probably limited by the reduced arterial pCO_2 (due to increased ventilation) since elevating the pCO_2 during hypoxia led to an additional dramatic increase in regional blood flow. However, after three weeks of hypoxia, cerebral blood flow had returned to values indistinguishable from that of the control group. LaManna *et al.* (1992) also measured the density of microvessels (predominantly capillaries) in nine regions of the brain from rats exposed to three weeks of hypobaric hypoxia and from littermate control rats. Microvessel density (number of microvessels per mm^2) was significantly higher in all nine brain regions in rats exposed to three weeks of hypoxia compared to control rats. As a consequence of this increase in capillary density, the diffusion distance for oxygen between capillary and brain cells (glia and neurons) would be shortened and shortening this diffusion distance would facilitate oxygen delivery to these cells. The time course of hypoxia induced angiogenesis has been described (LaManna *et al.*, 2004). Capillary density begins to increase by one week of hypoxia exposure and the process is completed between weeks 2 and 3 of exposure. The process is reversible. When rats previously exposed to hypoxia for three weeks are returned to a normoxic environment, there is a gradual decrease in capillary density back to baseline values. LaManna *et al.* (2004) have proposed that hypoxia inducible factor-1 (HIF-1) is a likely candidate for initiating angiogenesis. HIF-1 is an evolutionarily conserved transcription factor that functions as a main regulator of gene expression in response to hypoxia. HIF-1 activates many genes with hypoxic response elements. A more detailed discussion of the mechanisms underlying hypoxia induced brain angiogenesis is given in the paper by LaManna *et al.* (2004).

In summary, the initial responses to chronic mild hypoxia include an increase in cerebral blood flow and an increase in hematocrit. With persistence of the hypoxia, an increase in capillary density occurs which facilitates oxygen delivery to glia and neurons by reducing the diffusion distance between capillary and cell. These "global" adaptations are somehow integrated with the changes in regional blood flow that occur in response to neuronal activity since as reported by Lindauer *et al.* (2003) and discussed in the "Cerebral Blood Flow: Its Regulation" section, a low arterial pO_2 (around 45 mm Hg) augments the neuronal activated increase in regional blood flow.

Natural Animal Models

Before discussing natural hypoxia tolerant animal models, two key observations with respect to the neuronal response to severe hypoxia/ischemia are re-emphasized. Firstly, the critical initial event induced by hypoxia/ischemia is the dissipation of ion gradients with resultant membrane depolarization (anoxic depolarization). Once anoxic depolarization occurs, a number of secondary events are triggered which generally lead to cell death. Secondly, the time to onset of the anoxic depolarization appears to be a function of ATP levels. The longer ATP levels can be maintained, the more prolonged the time interval before the anoxic depolarization occurs. Hence, prevention of the anoxic depolarization and maintenance of cellular ATP levels would be two important strategies for protecting neurons from hypoxic/ischemic induced damage and ultimately cell death. These two basic neuroprotective strategies underlie the adaptations of hypoxia tolerant animals.

Turtle

Perhaps the best characterized hypoxia tolerant neurons are those from the freshwater turtle genus *Chrysemys*. The Western painted turtle can survive five months of anoxia at 1–3°C during winter dormancy (Bickler and Donohoe, 2002). In the following sections, what is known about the design features of the turtle brain which allows it to withstand prolonged anoxia are reviewed. Several other biochemical/physiologic characteristics of the turtle are important in determining the length of time this vertebrate can survive winter anoxic submergence (Warren *et al.*, 2006). During the anoxic period, anaerobic ATP production will generate lactic acid. Part of this lactic acid load is buffered by the turtle skeleton and shell. In addition, the painted turtle (*Chrysemys picta*) has been found to have an alkaline blood pH (about 8.00) and a blood bicarbonate concentration between 39–48 meq/l prior to winter anoxic submergence (Warren *et al.*, 2006). The painted turtle also has a large liver glycogen content which would serve as a source of glucose for the brain during the anoxic period. Hence, the turtle's capacity to buffer lactic acid and the size of tissue (liver) glycogen stores are important components underlying the anoxia tolerance of this vertebrate.

In the "General Features of Neuronal Response" section of, "Oxygen and Oxygen/Glucose Deprivation and the Brain", I discussed a report by Sick *et al.* (1982b) comparing the response of the rat and turtle to anoxia. There are other observations in this paper that were not discussed in that earlier section. Sick *et al.* (1982b) compared *in vivo* the response of the rat parietal cortex and the freshwater turtle (*Pseudemys scripta elegans*) telencephalon to anoxia induced by replacing inspired air with 100% nitrogen. Turtles were maintained at room temperature and rats were maintained at 37°C. In the rat, the mean extracellular brain potassium concentration increased rapidly after three minutes of anoxia to 77.7 mM. This abrupt rise in extracellular potassium correlates with the anoxic depolarization (Hansen, 1985). The rat brain also became isoelectric within one minute of the onset of anoxia. By contrast in the turtle, the mean extracellular brain concentration increased minimally from 2.59 to 3.68 mM after 30 minutes of anoxia and there was never an abrupt rise in extracellular potassium concentration even after four hours of anoxia. This observation was confirmed by Haddad and Jiang (1993a) who demonstrated that two hours of anoxia caused no change in turtle brain stem neuronal membrane potential. Sick *et al.* (1982b) also showed that the electrical activity of the turtle brain was suppressed by anoxia but the brain never became electrically silent even after four hours of anoxia. The different responses of the rat and turtle brains were not due to temperature effects since the rat brain showed the same responses to anoxia when the experiments were repeated at room temperature. Sick *et al.* (1982b) also examined the combined effects of anoxia and inhibition of anaerobic glycolysis on turtle brain function. Glycolysis was blocked by the addition of iodoacetate, an inhibitor of the enzyme glyceraldehyde-3-phosphate dehydrogenase. The combination of anoxia and iodoacetate resulted in an abrupt increase in mean extracellular potassium concentration to a value 20.6 mM. This rapid increase in extracellular potassium concentration occurred within 30 minutes of the onset of ischemia. However, unlike the situation in rat hippocampal CA1 neurons, in which the ischemic induced anoxic depolarization is not reversed by reintroduction of oxygen/glucose (Tanaka *et al.* 1997 and 1999), re-instatement of oxygen in the inspired gas resulted in reestablishment of control levels of extracellular potassium in the turtle brain even though iodoacetate was still present. This observation indicates that the turtle brain is very anoxia tolerant but not resistant to ischemia (oxygen/glucose deprivation) although the ischemic insult appears to be reversible with prompt reoxygenation.

The differential effects of anoxia and ischemia on turtle brain has also been examined by Doll *et al.* (1991). These investigators studied pyramidal neurons in brain slices prepared from rats and turtles (*Chrysemys picta*). Experiments were done at 25°C for both rats and turtles. Anoxia was induced by replacing oxygen in the medium with nitrogen and in some experiments (pharmacologic) anoxia was induced by adding cyanide. Ischemia was induced with iodoacetate combined with anoxia. Turtle neurons remained healthy throughout 180 minutes of anoxia which was the usual duration of the experiment. There was no change in membrane potential and these neurons were able to fire action potentials for the duration of the experiment. In several other studies, turtle neurons exposed to 18 hours of anoxia showed no changes in membrane potential. Rat cortical neurons exposed to anoxia depolarized to a membrane potential of 0 mV within a mean time of about 25 to 40 minutes depending upon whether the anoxia was induced by oxygen replacement with nitrogen or cyanide addition. Ischemia induced a rapid membrane depolarization in both turtle and rat neurons with the mean time between onset of ischemia and depolarization of the membrane to 0 mV being 4.6 minutes in the turtle and 3.1 minutes in the rat. The observation that ischemia in the turtle resulted in a rapid depolarization to 0 mV suggests that the turtle neurons were no longer viable. It is interesting that the effect of ischemia on the turtle neuron may also be dependent on the neuron population studied. Xia *et al.* (1992) measured the effect of ischemia (replacement of oxygen in the medium with nitrogen plus addition of iodoacetate) on potassium ion homeostasis in turtle brain stem slices (hypoglossal nucleus). Experiments were performed at 22–24°C. Mean extracellular potassium concentration increased by 5.4 mM above baseline value after 30 minutes of ischemia. This ischemia induced change in extracellular potassium concentration is much less than that reported by Sick *et al.* (1982b) and Chih *et al.* (1989a, reviewed next). These two sets of investigators used different brain regions; telencephalon in the study of Sick *et al.* and olfactory bulb in the study of Chih *et al.* Perhaps neurons of the hypoglossal nucleus in the turtle are even more anoxia tolerant than cortical neurons.

The study by Chih *et al.* (1989a) referred to in the preceding paragraph, addressed a number of questions related to the turtle's capacity to withstand prolonged anoxia. The brain (region of the olfactory bulb) of freshwater turtles (*Pseudemys scripta elegans*) was studied *in vivo* at 25°C. Anoxia was induced by replacing inspired gas with 100% nitrogen and ischemia was induced by combining anoxia with iodoacetate. There was no change in extracellular potassium concentration after six hours of anoxia but there was a decrease in amplitude of evoked potentials over the six-hour period. However, the amplitude of the evoked potentials recovered with reoxygenation. Ischemia (anoxia plus iodoacetate) produced a much different effect. There was initially a gradual increase in extracellular potassium concentration. However, within 35–90 minutes of the onset of ischemia, the extracellular potassium concentration rapidly increased to a mean value of 31.7 mM compared to a baseline value of 3.0 mM (mean value). Evoked potentials were significantly depressed in amplitude and became fully suppressed at the time of the rapid rise in extracellular potassium concentration (which as already noted would correspond to the anoxic depolarization). Interestingly, reoxygenation and removal of iodoacetate led to a return of extracellular potassium concentration to baseline within 90–130 minutes but there was no recovery of evoked potentials even after six hours of reoxygenation. This observation and the data of Doll *et al.* (1991) suggest that ischemia in the turtle causes irreversible damage even though ion gradients can be restored with reoxygenation (Sick *et al.*, 1982b) or with reoxygenation and removal of iodoacetate (Chih *et al.*, 1989a). Chih *et al.* (1989a) also measured ATP levels during anoxia and ischemia and as well calculated anaerobic ATP production during the course of anoxia from the accumulation of lactate and the dephosphorylation of creatine phosphate. ATP levels were unchanged after six hours of anoxia even though anaerobic ATP production had fallen to less than 10% of baseline value by four hours of anoxia. During ischemia, ATP levels did fall and were 67% of baseline just prior to the anoxic depolarization and 54% of baseline immediately after the anoxic depolarization. The observations of Chih *et al.* (1989a) clearly point out that anoxic depolarization does not occur when ATP levels are maintained close to baseline levels as occurs during anoxia. However, these investigators also noted that the fall in ATP levels during ischemia was relatively small when a comparison is made between values just before and just after the anoxic depolarization. This small fall in the level of ATP is unlikely to decrease the transport activity of the Na^+-K^+-ATPase pump sufficient to explain the experimentally measured rate of increase in the extracellular potassium concentration associated with the anoxic depolarization. The rapid loss of ion gradients leading to the anoxic depolarization is more likely due to changes in membrane conductance rather than simply failure of Na^+-K^+-ATPase pump activity. This conclusion is similar to the one reached by Hansen (1985) with respect to the mammalian brain (see "General Features of Neuronal Response" section in "Oxygen and Oxygen/Glucose Deprivation and the Brain"). The study by Chih *et al.* (1989a) also indicates that the maintenance of ATP levels during anoxia by the turtle neuron is due to a reduction in utilization and not due to an increase in anaerobic ATP production.

In the mammalian brain, hypoxia or ischemia induced depolarization triggers a series of secondary events which include an increase in extracellular glutamate concentration and an increase in the concentration of free intracellular calcium. Given that hypoxia does not induce depolarization in the turtle brain, one would expect such secondary events not to occur.

Indeed that is the case. Young *et al.* (1993) used brain stem slices from rats and turtles (*Pseudemys scripta elegans*) to study the effects of anoxia on the extracellular concentration of excitatory amino acids. Experiments for both rats and turtles were performed at 36°C and anoxia was induced by replacing oxygen in the medium with nitrogen. Ten minutes of anoxia in the rat brain stem slice resulted in a four fold increase in extracellular aspartate concentration and a two fold increase in extracellular glutamate concentration compared to baseline levels. In the turtle brain stem slices, 60 minutes of anoxia produced only a 50% increase in extracellular aspartate concentration compared to baseline levels and in the case of glutamate there was actually a decrease in extracellular concentration during the anoxic period. With respect to calcium homeostasis, Bickler and Buck (1998) reported that the free intracellular calcium concentration in cortical sheets from the turtle *Chrysemys picta* increased by about 10% (135 nmol/l to 150 nmol/l) after 40 days of anoxia at 3°C. Bickler *et al.* (2000) measured the free intracellular calcium concentration in cortical sheets from the turtle (*Chrysemys picta*) after two hours of anoxia (induced by replacement of oxygen in the medium with nitrogen). Measurements were made at 25°C. Mean calcium concentration rose from a baseline of 135 nmol/l to 202 nmol/l at the end of the two-hour anoxic period. In adult rat cortical brain slices, 20 minutes of anoxia (replacement of oxygen in the medium with nitrogen) resulted in an increase in free intracellular calcium concentration from 165 nmol/l to about 1800 nmol/l (Bickler *et al.*, 1993). These experiments were performed at 37°C. Similarily, Bickler and Hansen (1994) showed that in rat cortical brain slices, ten minutes of anoxia (replacement of oxygen in the medium with nitrogen) caused the free intracellular calcium concentration to rise from a baseline value of 138 nmol/l to about 1100 nmol/l. These experiments were done at 30–31°C. Hence during prolonged anoxia, the turtle brain unlike the rat brain, demonstrates only a modest increase in free intracellular calcium concentration and no increase in the extracellular concentration of glutamate.

To summarize, the turtle brain is very tolerant of prolonged periods of anoxia. ATP levels are maintained at normal levels despite reduced anaerobic ATP production. Clearly ATP utilization must be significantly decreased and a new balance is achieved between levels of utilization and production. Ion gradients are maintained and anoxic depolarization does not occur. Brain electrical activity is suppressed but not eliminated and full recovery occurs with reoxygenation. In contrast, the turtle brain, similar to the mammalian brain, is sensitive to ischemia. ATP levels fall, ion gradients are dissipated and anoxic depolarization occurs. Brain electrical activity (evoked potentials) is completely suppressed and not recovered with removal of the ischemic insult indicating that irreversible cell damage has occurred. The key features then, of the turtle brain's anoxia tolerance are maintenance of ATP levels through reduced utilization and prevention of anoxic depolarization. These observations lead to two inter-related questions: what are the underlying mechanisms by which the turtle reduces ATP utilization and what are the mechanisms by which the turtle prevents anoxia induced depolarization?

Reduction of ATP utilization by the turtle

The metabolic characteristics of the turtle brain have been examined by a number of investigators. Suarez et al. (1989) estimated glucose utilization rates in the turtle (Chrysemys picta) brain under normoxic conditions at 25°C. They reported values about 1/12 the rate for rat brain. They also compared brain hexokinase, lactic dehydrogenase and citrate synthase activities between turtle brain (25°C) and rat brain (37°C). Hexokinase and lactic dehydrogenase activities were comparable in the turtle and rat but citrate synthase activity was much higher in the rat. These results are consistent with the observations of Xia et al. (1992). Turtle (Pseudemys scripta elegans) brain and rat brain had comparable hexokinase activity, but cytochrome c oxidase activity was higher in the rat. The temperature at which these measurements were performed was not given. Hence, turtles and rats have similar glycolytic capacities but the rat has a higher oxidative capacity. This data indicate that the metabolic machinery of the turtle brain is not designed to compensate for decreased ATP production during anoxia by an increase in anaerobic ATP production. In fact, the lower glucose utilization rate and aerobic capacity characteristic of the turtle brain indicate that ATP turnover (production and utilization) occurs at a much lower rate in turtle brain (compared to rat brain) even under normoxic conditions. Suarez et al. (1989) also measured Na^+-K^+-ATPase activity in rat and turtle brain (assayed at 25°C in both species) and noted that Na^+-K^+-ATPase activity was about 2–2.5 times higher in rat than turtle brain. In addition, to lower Na^+-K^+-ATPase activity (Suarez et al. 1989), the turtle brain also has a 70% lower density of voltage dependent sodium channels than the rat brain (Lutz and Milton, 2004). These observations suggest that at least one mechanism underlying the decrease in ATP utilization rate (of the turtle brain) is an alteration in membrane transport processes. Is there evidence that membrane transport processes are modified in the turtle?

Despite the large reduction in anaerobic ATP production during anoxia (Chih *et al.*, 1989a), the turtle brain is able to maintain ion gradients and does not undergo (anoxic) depolarization. This situation is remarkable given that a substantial amount of energy in mammalian cells is normally expended on ion transport pumps (Bickler and Buck, 1998). The combination of decreased energy production yet maintenance of ion gradients suggests that an inherently low ion permeability exists in hypoxia tolerant cells such as the turtle neuron. The low ion permeability of hypoxia tolerant compared to hypoxia sensitive neurons would mean that less energy would have to be consumed maintaining ion gradients through the action of active ion pumps. Hence, the ATP utilization rate would be reduced and if reduced enough a balance could be achieved between this lowered ATP utilization rate and the decreased anaerobic ATP production rate characteristic of the anoxic state. These observations have been incorporated into a more formal "channel arrest" hypothesis (Hochachka and Lutz, 2001) which comprises two postulates: (1) hypoxia tolerant cells would have an inherent low ion permeability (low channel densities or low channel activities) and (2) during anoxia a further suppression of membrane ion permeability would occur.

There is evidence for postulate No. 1 of this hypothesis. Doll *et al.* (1993) compared whole cell (Gw) and specific membrane (Gm) conductance of turtle (*Chrysemys picta*) and rat pyramidal neurons in slices. Anoxia was induced by replacing oxygen in the medium with nitrogen. Under normoxic conditions, turtle neuron Gw was much less than that of the rat. The ratio of rat to turtle Gw varied with the temperature at which the measurements were made. At 25°C for both species, the ratio (rat/turtle) of Gw was 4.2. At 37°C for the rat and 15°C for the turtle, the ratio (rat/turtle) for Gw was 22. When turtle neurons were subjected to anoxia (at 25°C) for six to nine hours, there was no change in membrane potential and no further reduction in either Gw or Gm from baseline values. This observation implies that no additional suppression of membrane ion permeability occurred during anoxia. More recently, Ghai and Buck (1999) reported a 27% reduction in Gw in turtle (*Chrysemys picta*) cortical brain sheets during anoxia (induced by replacing oxygen with nitrogen in the medium). These investigators did not include EGTA in the recording electrode and they suggested that inclusion of EGTA in the recording electrode was the most likely reason an anoxia induced reduction in Gw was not detected by Doll *et al.* (1993). The mechanism by which EGTA in the recording electrode might block an anoxia induced decrease in Gw is unclear. Ghai and Buck (1999) also observed that bathing the cortical brain sheets in a high calcium medium resulted in a reduction in Gw under normoxic conditions and that this reduction occurred even if EGTA was included in the recording electrode. Hence the data of Ghai and Buck (1989) would be evidence in support of postulate No. 2 of the "channel arrest" hypothesis. Chih *et al.* (1989b) measured potassium ion leakage in turtle neurons during anoxia as a further test of postulate No. 2. In this *in vivo* study, extracellular potassium concentrations were measured in normoxic and anoxic turtle (*Pseudemys scripta elegans*) brains superfused with oubain to block the Na^+-K^+-ATPase pump. All experiments were conducted at 25°C. Anoxia was induced by having the turtles inspire 100% nitrogen. The investigators reasoned that the efflux of potassium after the Na/K pump was blocked would be a measure of passive potassium movement through leakage channels. The time required for the extracellular potassium concentration to increase by 1 mM above resting levels in normoxic brains was about 51% of that in anoxic brains, indicating that the rate of potassium leakage in normoxic brains was almost twice that in anoxic brains. This observation would be consistent with an anoxia induced decrease in membrane ion permeability and evidence in support of postulate No. 2. However, if in the turtle brain the anoxia induced reduction in ion permeability is mainly due to inhibition of potassium leak channels then this effect should change the membrane potential. As discussed with respect to the mammalian brain, hypoxic inhibition of potassium "leak" channels leads to membrane depolarization (Plant *et al.*, 2002). Yet, no change in membrane potential occurs in turtle neurons even after prolonged anoxic exposure. There is some evidence that sodium channel density is decreased in the turtle cerebellum during anoxia (Lutz and Milton, 2004) and this change in sodium channel density may be a part of the mechanism by which the turtle suppresses brain electrical activity during anoxia (Lutz and Milton, 2004; Sick *et al.*, 1982b; Chih *et al.*, 1989a). Neuronal activity in the mammalian brain is associated with significant increases in the extracellular concentration of potassium and decreases in the extracellular concentration of calcium (Hansen, 1985) and these changes in ion concentrations would have to be restored by active pumping mechanisms. Hence, important energy savings can be made by suppression of brain electrical activity during anoxia and this may be an important strategy by which the turtle brain conserves ATP. Perhaps the most direct evidence for postulate No. 2 of the "channel arrest" hypothesis is the effect of anoxia on NMDA receptors (NMDAR) in the turtle. Bickler *et al.* (2000) studied these receptors in cerebrocortical sheet pyramidal neurons of the turtle *Chrysemys picta*. Anoxia was induced by replacing oxygen in the perfusing medium with nitrogen and these studies were performed at 25°C. In neurons exposed to three hours of anoxia, membrane potential remained stable and mean intracellular ATP level remained within 80% of the mean control value. In cell attached patches, within 15 minutes of the onset of anoxia, the open probability of NMDAR was reduced by about 50%. No change in single-channel receptor current amplitude was observed during anoxia. This rapid suppression of receptor activity within minutes of the onset of anoxia was due to a change in receptor protein phosphorylation state. Within three days of the onset of anoxia, the abundance of NMDAR decreased by approximately 60% and this reduced abundance persisted for the duration of the experiment (21 days of anoxia). Within five to seven hours of reoxygenation, receptor abundance returned to the baseline level. The survival advantage of these anoxia induced changes in NMDAR (reduced open probability within minutes and reduced receptor abundance over days) is not completely clear to me. During anoxia, the turtle brain does not depolarize and the extracellular glutamate concentration actually decreases. Hence both of these factors would mitigate against any increase in the function NMDAR during anoxia. Perhaps the hypoxia induced silencing of NMDAR during anoxia represents another mechanism for saving energy; in this case by a reduction in excitatory neurotransmission.

Hence in summary, during anoxia there is suppression of both neuronal electrical activity and NMDAR function in the turtle brain and these changes would account for at least part of the reduction in ATP utilization. The evidence is less strong that anoxia induces reduced activity of potassium leak channels since no change occurs in membrane potential during anoxia.

Prevention of anoxia induced depolarization

As is true of the mammalian brain, the characteristics of the turtle brain point to a clear link between intracellular ATP levels and whether or not anoxia induces membrane depolarization. During anoxia, the turtle brain is able to maintain a near normal intracellular ATP concentration and membrane depolarization does not occur. If the turtle brain is subjected to ischemia, the intracellular ATP concentration falls and membrane depolarization occurs. However, the nature of the link between intracellular ATP concentration and the depolarization process is unclear. As has been noted for both the mammalian and turtle brain, the link is not simply due to failure of active ion pumps to maintain ion gradients. A large increase in ion conductance probably initiates the anoxic depolarization. The concomitant failure of active ion pumps such as the Na^+-K^+-ATPase pump prevents the restoration of these ion gradients. However, the turtle data does tell us that a low tissue pO_2 alone, either directly or indirectly, is not a sufficient stimulus to activate the large conductance channels responsible for the dissipation of ion gradients and generation of the membrane depolarization. The process initiating the anoxic depolarization is energy linked and is somehow triggered by either a low concentration of ATP or possibly by a combined effect of a low tissue pO_2 and a low concentration of ATP. As already discussed with respect to the mammalian brain, Thompson *et al.* (2006) showed that ischemia (oxygen/glucose deprivation) opens large hemichannels formed from unopposed half gap junctions although recent observations by Douglas *et al.* (2006) indicate that hemichannel opening does not appear to be the primary initiating process generating the anoxic depolarization. Interestingly, Ghai and Buck (1999) have made the observation that in the turtle brain, gap junctions may provide a mechanism whereby anoxia can alter neuronal whole cell conductance. These investigators speculated that anoxia closes gap junctions in the turtle brain and thus decreases neuronal whole cell conductance. Since the actual channels involved in initiating the anoxic depolarization are unknown, there are no data as to how the turtle neuron actually prevents the depolarization process when exposed to anoxic conditions. However, there are limited data by which one can compare the relative size of the ischemic channels in rat and turtle. For example, Hansen (1985) reported that in the rat the extracellular potassium concentration during the anoxic depolarization increases at a rate of 80 mM per second. By comparison, in the turtle brain rendered ischemic, the extracellular potassium concentration increases at a rate of approximately 26 mM per minute during the anoxic depolarization (Chih *et al.*, 1989a). This crude comparison suggests that the channels responsible for ischemic induced depolarization in the turtle brain are much smaller than those of the rat brain.

In summary, the chief biochemical/physiologic characteristics underlying the ability of the turtle neuron to withstand prolonged periods of anoxia are the following: (1) The capacity to maintain intracellular ATP levels close to baseline through a significant reduction in ATP utilizing processes. (2) Prevention of anoxic depolarization. (3) Maintenance of calcium homeostasis with only modest increases in intracellular calcium concentration. However, since the rise in intracellular calcium concentration in the mammalian brain during anoxia is a secondary event triggered by the anoxic depolarization, the reason the anoxic turtle brain can maintain calcium homeostasis is most likely because anoxic depolarization does not occur. (4) Reduction in the activity of NMDA receptors. It is interesting that these same biochemical/physiologic characteristics are present, although to a lesser extent, in hypoxia tolerant mammalian neurons such as those of the neonate. The turtle brain reduces ATP utilization by suppressing neuronal electrical activity and perhaps by other modifications in membrane ion transport processes.

Crucian carp (*Carassius carassius*)

The crucian carp is an extremely hypoxia tolerant fish which can survive anoxia for weeks at 10°C and for months at temperatures close to zero. The metabolic characteristics of this fish are described in several recent publications (Nilsson and Renshaw, 2004; Nilsson, 2001). At the onset of anoxia, this fish starts producing ethanol as the major end product of anaerobic glycolysis. Ethanol production only occurs in muscle and during anoxia all other tissues produce lactate which is transported by the circulation to the muscle. There the lactate is converted to pyruvate which is transformed to acetaldehyde and subsequently to ethanol. The brain of the crucian carp demonstrates similar biochemical/physiologic characteristics to that of the turtle. Johansson and Nilsson (1995) studied ion homeostasis *in vivo* in the carp telencephalon. Measurements were made at 10°C and anoxia was initiated and maintained by bubbling nitrogen into the respiratory water. Crucian carp brain showed no increase in extracellular potassium concentration even after 15 hours of anoxia. After two hours of anoxia, the intracellular concentration of ATP in the telencephalon was unchanged from the baseline value. However, the combination of anoxia and inhibition of glycolysis with iodoacetate resulted in a sudden rapid rise in the extracellular potassium concentration after about 30 minutes. This rapid increase in extracellular potassium concentration would correspond to the anoxic depolarization. During the depolarization, the rate of rise of extracellular potassium was about 0.5 mM per minute which is a much lower rate than that of the turtle brain (see previous section). ATP concentration fell in the ischemic carp brain and at the start of the anoxic depolarization was about 6.9% of baseline level. There are several other comparable features between carp and turtle brains. The extracellular glutamate concentration does not increase in the carp brain even after five hours of anoxia (Nilsson and Renshaw, 2004) and apparently the anoxic carp brain does not experience a significant rise in intracellular calcium concentration (Nilsson, 2001). Hence, three of the chief biochemical/physiologic characteristics of the hypoxia tolerant turtle brain (see previous section) have been described in the carp brain. Presumably, the anoxic carp brain maintains a normal ATP concentration by decreasing ATP utilization since there is at least a 30%–40% reduction in ATP turnover during anoxia (Nilsson and Renshaw, 2004). Brain electrical activity is suppressed during anoxia (Nilsson, 2001) which would reduce ATP consumption as discussed with respect to the turtle brain.

High altitude birds

Birds are generally more tolerant of anoxia than mammals. For example, house sparrows at an altitude of 6100 meters are alert, active and exhibit normal behavior whereas mice exposed to the same altitude are comatose (Faraci, 1991). Bar-headed geese (*Anser indicus*) fly over the Himalayan mountains twice a year on their migratory route between Tibet and India and have been sighted flying above the summit of Mount Everest, 8848 meters above sea level (Scott and Milsom, 2006). Most of the studies examining the hypoxia tolerance of high altitude birds have focused on adaptations designed to maximize tissue oxygen delivery. There are no known studies on the biochemical/physiologic responses of the brain of high-altitude birds to anoxia or ischemia comparable to studies done in the mammal, turtle and carp.

The bird lung is unique amongst the lungs of air breathing vertebrates with a blood flow that is crosscurrent to gas flow and a gas flow that occurs unidirectionally through rigid parabronchioles (Scott and Milsom, 2006). These features make the lungs of birds more efficient with gas exchange than the lungs of other vertebrates. However, these features are present in all birds and so cannot be considered adaptations specific to high altitude fliers. The ventilatory response to hypoxia (primarily due to an increase in breathing frequency) is similar between birds that do and do not fly at high altitudes. Hyperventilation results in a respiratory alkalosis as well as hypocapnia. If high altitude birds were less sensitive to these changes in CO_2/pH compared to low altitude birds, they might be better able to sustain increases in ventilation during severe hypoxia coupled with exercise. However, whether high versus low altitude birds have a different sensitivity to the ventilatory induced changes in CO_2/pH has never been compared (Scott and Milsom, 2006).

Birds differ from mammals in terms of regulation of the cerebral circulation. A decrease in pCO_2 in mammals causes vasoconstriction and a reduction in cerebral blood flow. In bar-headed geese, reductions in arterial pCO_2 to less than 10 mmHg do not change cerebral blood flow (Faraci, 1991). The mechanism underlying the reduced sensitivity of avian cerebral vessels to hypocapnia is unknown. However, the ability to maintain cerebral blood flow during hypocapnia/alkalosis is thought to be present in all birds and not just those that fly at extreme altitudes (Scott and Milsom, 2006).

One adaptation that seems to be specific to high altitude species is an increase in the affinity of their hemoglobin for oxygen. Liang *et al.*, (2001) compared the structure of oxyhemoglobin in the greylag and bar-headed goose. The greylag goose (*Anser anser*) lives on lowlands and does not tolerate hypoxic conditions whereas the bar-headed goose, as already noted, flies at extreme altitudes and is very hypoxia tolerant. Liang *et al.* (2001) found four amino acid residue differences between the hemoglobins of these two species. These differences result in a unique quaternary structure for the bar-headed goose oxyhemoglobin compared to the quaternary structure of the oxyhemoglobin of the greylag goose, chicken and human which all have a similar quaternary structure. These differences in quaternary structure are responsible for differences in oxygen affinity. For example, the P50 (pO_2 at half saturation of hemoglobin with oxygen) a measure of oxygen affinity, is 42.6 mmHg for the Pekin duck (hypoxia intolerant) but 27.2 mmHg for the bar-headed goose and 16.4 mmHg for the vulture *Gyps rueppellii*, both species being high flyers (Faraci, 1991). However, although a lower P50 will increase oxygen loading at the lungs, it will inhibit oxygen unloading at the tissues. Since the Bohr (CO_2/pH) effect on hemoglobin oxygen affinity is less in bar-headed geese than in other waterfowl (Scott and Milsom, 2006), changes in other (as yet undefined) factors at the tissue level must facilitate oxygen unloading.

Another reported adaptation is a greater capillary density in the pectoral muscle of the bar-headed goose compared to the low altitude pekin duck (Scott and Milsom, 2006). Such an increase in capillary density would reduce oxygen diffusion distance and hence facilitate delivery of oxygen to cells. A similar adaptation was noted in the brain of rats exposed to chronic hypobaric hypoxia as described in the "Chronic Mild Hypoxia and the Brain" section.

In an attempt to quantify the importance of various physiological adaptations for high altitude bird flight, Scott and Milsom (2006) performed a sensitivity analysis of oxygen consumption using a number of respiratory, circulatory and tissue traits. They modeled oxygen uptake by the capillaries in the lung and oxygen transport at the tissue level. They used data from the literature on pekin ducks when available to reproduce *in vivo* conditions under three sets of circumstances: normoxia (sea level, inspired oxygen tension 150 mmHg), moderate hypoxia (3500 meters, inspired oxygen tension 84 mmHg) and severe hypoxia (10,000 meters, inspired oxygen tension 30 mmHg). According to the analysis of Scott and Milsom (2006), the following variables appear to be the most important with respect to maximizing oxygen consumption during flight at an altitude of 10,000 meter (severe hypoxia). (1) Hemoglobin affinity: decreasing P50 from 40 to 10 mmHg increased oxygen consumption by about 20%. (2) Oxygen diffusion capacity of the tissues (DtO_2): there was a significant interaction between P50 and DtO_2. Doubling DtO_2 at a P50 of 40 mmHg resulted in only a slight increase in oxygen consumption (2%), whereas a doubling of DtO_2 at a P50 of 25 mmHg increased oxygen consumption by 34%. At a lower P50, there is a substantially higher arterial oxygen content so more oxygen can be removed and increasing DtO_2 will have a greater influence. Tissue diffusion capacity can be augmented by a greater tissue capillarization which has been demonstrated in the pectoral muscle of the bar-headed goose. (3) Total ventilation: at a P50 of 40 mmHg doubling total ventilation increased oxygen consumption by 41%, whereas at a P50 of 25 mmHg, doubling total ventilation increased oxygen consumption by 33%. (4) Temperature: hyperventilation during flight at extreme altitude would cool the pulmonary blood and reduce P50 of hemoglobin in the lungs and thus facilitate oxygen uptake. When this blood enters tissues (such as exercising muscle and brain) it would be rewarmed to body temperature and oxygen would be released from hemoglobin.

It is interesting that the analysis of Scott and Milsom (2006) indicated that an increase in hemoglobin concentration did not make a big impact on oxygen consumption at a high altitude (severe hypoxia). Doubling the hemoglobin concentration increased oxygen consumption by 10% at a P50 of 40 mmHg and by only 2% at a P50 of 25 mmHg.

Thus the high altitude bird maximizes oxygen consumption (energy flux) primarily through a combination of hyperventilation, high affinity hemoglobin and an increase in tissue oxygen diffusion capacity. It would be extremely interesting to know if the brain of these high fliers is also adapted to withstand an extremely hypoxic environment. Unfortunately no data are available on this particular topic.

Concluding Remarks and Future Directions

Since the brain is extremely sensitive to deprivation of oxygen or oxygen/glucose, this chapter is essentially limited to a discussion of the effects of anoxia (oxygen deprivation) and ischemia (oxygen/glucose deprivation) on brain function. Even with this limitation, the number of published studies on this topic is extensive and in this chapter only a fraction is reviewed. As emphasized, the initial critical events induced by anoxia/ischemia in the mammalian brain are a rapid dissipation of ion gradients with membrane depolarization (termed anoxic depolarization). This anoxic depolarization triggers a series of secondary events which include among others an increase in free intracellular calcium concentration, an increase in the extracellular concentration of excitatory neurotransmitters such as glutamate and activation of glutamate receptors. It is clear though, that the seminal event is the anoxic depolarization and that the primary strategy used by the turtle and carp brain to avoid hypoxic damage is to prevent the anoxic depolarization. Yet, as one reads the published literature it is apparent that the majority of studies on anoxic/ischemic brain damage have focused on the "secondary" events and not on the mechanisms underlying the anoxic depolarization. For example, in a recent review article, Barber *et al.* (2003) only devote part of a sentence to the occurrence of membrane depolarization. As discussed in the "General Features of Neuronal Response" section, the mechanisms underlying the anoxic depolarization in the mammalian brain are still not well understood. Operationally, the anoxic depolarization can be described as comprising two processes: the opening of large conductance channels coupled with failure of active ion pumps such as the Na^+-K^+-ATPase pump. As a result of these large conductance channels opening, ion gradients dissipate and the membrane depolarizes. Which channels are responsible for this event is not known, but as previously discussed, several types of "megachannels" have been proposed as possible candidates. How the opening of these large conductance "megachannels" is coupled to the failure of active ion pumps is unclear, but the linkage involves the adequacy of intracellular ATP stores. As a result of oxygen or oxygen/glucose deprivation, ATP production falls, intracellular ATP stores become depleted and consequently active ion pumps fail. As a manifestation of this coupling, the onset of anoxic depolarization is delayed the longer adequate levels of intracellular ATP can be maintained. In the case of the turtle and carp, anoxic depolarization does not occur at all during prolonged periods of anoxia, because ATP levels in the brain of these vertebrates are maintained close to normal (i.e. levels found during normoxic conditions). However, it should be noted that anoxic depolarization is not simply due to active ion pump failure; the opening of large conductance "megachannels" must also occur. What signals trigger the opening of these channels is unknown. Anoxia alone does not appear to be a sufficient signal given the observations in the turtle and carp that prolonged anoxia does not result in anoxic depolarization as long as intracellular ATP stores remain close to normoxic levels. Perhaps these channels are opened (activated) by anoxia in combination with "low" intracellular ATP concentrations, although this postulate is purely speculative on my part.

In addition, developing a clearer picture of the mechanisms responsible for the anoxic depolarization has been hampered by several experimental variables which make comparisons between studies difficult. Firstly, not all mammalian neurons are equally anoxia sensitive. Neonatal neurons and probably also dorsal vagal neurons are more anoxia tolerant than mature hippocampal CA1 neurons. Hence, the response of a given population of neurons to anoxia/ischemia is not necessarily generalizable to all mammalian neurons. Secondly, tissue oxygen levels are often not well documented making it difficult to compare the severity of the anoxic insult between different studies. Finally, the term anoxic depolarization appears to be used rather loosely to describe any membrane depolarization occurring in the setting of anoxia/ischemia even though the underlying mechanism(s) are clearly different. For example, the composite features of the anoxic depolarization as described in the rat parietal cortex *in vivo* by Sick *et al.* (1982b) and Hansen (1985) and in the rat hippocampus *in vitro* by Thompson *et al.* (2006) and Tanaka *et al.* (1997) consist of a rapid massive depolarization accompanied by a large increase in conductance and a rapid increase in the extracellular concentration of potassium and a rapid decrease in the extracellular concentrations of sodium, chloride and calcium. By contrast the anoxic depolarization in cerebellar granule neurons *in vitro* described by Plant *et al.* (2002) consists of a rapid depolarization (of about 14 mV) accompanied by an increase in membrane resistance (decrease in conductance) and a decrease in outward current. In the first example, the underlying mechanism involves the sudden opening of large conductance channels with Thompson *et al.* (2006) proposing in their experiments that these large conductance channels are hemichannels formed from unopposed half gap junctions. In the second case, the depolarization and decrease in membrane conductance is due to the anoxia induced closing of potassium leak channels. Although in both cases the depolarization has been termed anoxic depolarization, the underlying mechanisms are quite different.

These comments will serve as a background to the next section which contains a synopsis of early events in the responses of the mammalian and turtle/carp brains to anoxia and ischemia.

In this synopsis, anoxia and ischemia will be considered parts of a continuum with respect to the production of ATP. Anoxia (oxygen deprivation) results in the inhibition of aerobic production of ATP, whereas during ischemia (oxygen and glucose deprivation) both the anaerobic and aerobic production of ATP are inhibited.

Mammalian brain

Anoxia

The mammalian brain displays a number of responses to anoxia which probably reflect both the differences in hypoxia tolerance of different neuron populations as well as the duration and severity of the oxygen deprivation. An early response to anoxia is membrane hyperpolarization which is generally transient but can be persistent as in the DVN. This hyperpolarization is mediated by the opening of ATP sensitive potassium channels as a result of low intracellular ATP levels. This membrane hyperpolarization would reduce neuronal activity by raising the action potential threshold and hence might represent a mechanism whereby ATP utilization by active ion pumps is decreased — a protective response. In the majority of mammalian neurons, the initial hyperpolarization is transient and with persistence of the anoxia, membrane depolarization occurs. As already discussed, a number of mechanisms have been implicated as underlying the depolarization process. Jiang and Haddad (1994) and Plant *et al.* (2002) have suggested that the depolarization is due to closing of oxygen sensing potassium channels. In fact, Jiang and Haddad have suggested that the potassium channels mediating the initial hyperpolarization are both ATP and oxygen sensitive and that with persistent anoxia these channels are closed by a direct effect of a low oxygen tension on the channel itself (or possibly the surrounding membrane). Another mechanism that could be responsible for depolarization is the opening of oxygen-sensing sodium channels as proposed by Hammarstrom and Gage (2000). These investigators also suggested that the effect of anoxia to open these channels may be irreversible since this effect of hypoxia was not readily reversed by reoxygenation. The consequences of membrane depolarization would be an increase in intracellular calcium concentration as well as an increase in the extracellular concentration of glutamate. However, if the anoxic insult is sufficiently prolonged and severe, intracellular ATP levels will fall to a critically low level (as yet undefined) which will trigger the opening of large conductance "megachannels." Ion gradients will be dissipated, the intracellular calcium concentration will increase steeply and a cascade of events will occur leading to cell death. Throughout the whole anoxic episode, as intracellular ATP levels fall, active ion pumps will begin to fail. However, in general, the process does not become irreversible until intracellular ATP levels fall to some as yet undefined critically low level which somehow triggers the opening of the large conductance "megachannels."

Ischemia

During ischemia, intracellular ATP depletion is much more profound then during anoxia and consequently failure of active ion pumps will be more complete. When exposed to ischemic conditions, mammalian neurons invariably display a rapid massive depolarization which may be preceded by a transient hyperpolarization and a slow depolarization phase (Tanaka *et al.* 1997). The hyperpolarization is probably due to the opening of ATP sensitive potassium channels, and the slow depolarization phase is probably a consequence of the failing Na^+-K^+-ATPase pump. The rapid massive depolarization is due to the opening of large conductance "megachannels" which occurs when intracellular ATP levels reach a critically low level. Since ischemia results in the inhibition of both aerobic and anaerobic ATP production, intracellular ATP levels will fall more precipitously during ischemia than during anoxia. With the opening of these large channels, ion gradients will dissipate and the intracellular concentration of calcium will increase often to a greater extent than observed during anoxia. When the rapid massive depolarization occurs, the condition of the neuron becomes one of irreversible damage. Presumably during the course of the ischemic insult, oxygen-sensing sodium channels will be opened and oxygen-sensing potassium channels will be closed, although at what time points during ischemia these events might occur is unclear. During anoxia, the closing of oxygen-sensing potassium channels and/or the opening of oxygen-sensing sodium channels may be the primary mechanisms mediating membrane depolarization if the anoxic insult has not been prolonged or severe enough to decrease intracellular ATP levels to the critically low level which triggers the opening of the large conductance "megachannels." During ischemia, however, since the inhibition of ATP production is more severe than during anoxia, the reduction of ATP levels to the critically low value that triggers the opening of the large conductance "megachannels" will, for all intents and purposes, always occur. This distinction between anoxia and ischemia is important and warrants repeating. If the anoxic insult is not severe enough to decrease ATP levels to the critically low (as yet undefined) level that triggers the opening of the large conductance "megachannels," then membrane depolarization will be mediated primarily by the closing of oxygen-sensing potassium channels and/or the opening of oxygen-sensing sodium channels. The membrane depolarization due to the closing of oxygen-sensing potassium channels is reversible (Plant *et al.*, 2002) and it is still not clear whether the depolarization due to the opening of oxygen-sensing sodium channels is or is not reversible (Hammarstrom and Gage, 2000). On the other hand, ischemia invariably leads to such a profound depletion of intracellular ATP that the opening of large conductance "megachannels" always occurs. Furthermore, it is probably reasonable to assume that the rapid massive depolarization that occurs when anoxia is severe and the massive rapid depolarization that occurs during the course of ischemia are comparable events and actually reflect the opening of the same "megachannels."

Turtle/carp brain

Anoxia

The turtle and carp brain can withstand prolonged periods of severe anoxia without displaying any change in neuronal membrane potential. This observation implies that oxygen-sensing potassium and sodium channels comparable to those found in mammalian neurons are not present in the turtle or carp neuron. In fact, under baseline normoxic conditions turtle neurons have a much lower membrane conductance than rat neurons (Doll *et al.*, 1993) indicating that baseline neuronal ion leak rates are much smaller in the turtle than in the rat. This observation is consistent with the finding that Na^+-K^+-ATPase activity is about 2–2.5 fold higher in the rat brain than in the turtle brain (Suarez *et al.*, 1989). Although anoxia severely reduces ATP production in the turtle brain (and presumably in the carp brain), intra-cellular ATP levels remain essentially unchanged from values measured under normoxic conditions. As a result of this maintenance of normal intra-cellular ATP levels, the opening of large conductance "megachannels" is prevented. The turtle brain is able to maintain intracellular ATP levels by drastically reducing ATP utilization as has been discussed in a previous section. In the mammalian brain, anoxia induced membrane depolariza-tion triggers a series of secondary events which include an increase in extracellular glutamate concentration and an increase in intracellular cal-cium concentration. However in the turtle and carp brain, since anoxia does not induce membrane depolarization, extracellular glutamate concen-tration remains low and increases in intracellular calcium concentration are very modest. It is also interesting that the biochemical/physiologic features underlying the relative anoxia tolerance of mammalian neonatal neurons are similar to those of the turtle and carp. Thus the turtle and carp do not appear to have a unique set of characteristics responsible for their extreme anoxia tolerance. Rather these vertebrates appear to have "fine tuned and extended" characteristics that are probably present in most (all?) neurons.

Ischemia

Although the turtle and carp brain are extremely hypoxia tolerant, they are sensitive to ischemia. The brain of both of these vertebrates demonstrates a rapid massive depolarization when subjected to an ischemic insult. In addition, intracellular ATP levels fall during the course of ischemia. Chih *et al.* (1989a) reported that in the turtle brain the intracellular ATP concen-tration had fallen by about 46% just immediately after the ischemia induced depolarization. In the carp brain, Johansson and Nilsson (1995) found that the intracellular ATP concentration was about 6.9% of baseline value at the start of the ischemia induced depolarization. In addition, in the turtle brain, Chih *et al.* (1989a) observed that evoked potentials which were fully suppressed at the time of the depolarization did not recover even after six hours of reoxygenation. Hence, the ischemia induced depolarization in the turtle brain results in irreversible neuronal damage as it does in the mam-mal. If one extrapolates from the mammalian data, then most likely the rapid depolarization that occurs during ischemia in turtle and carp neurons is due to the opening of large conductance channels although the existence of such channels has never been documented. As discussed in a previous section, if one uses the rate of rise of the extracellular potassium concentration during the ischemic depolarization as a crude measure of the size of these large conductance channels, then these channels are much smaller in turtle and carp neurons than in rat neurons. The observations made in the ischemic turtle and carp brain support the notion that reduced intracellular ATP levels are a prerequisite to the opening of large conductance channels.

In this chapter, the main focus has been on the early events following the exposure of mammalian neurons to an anoxic or ischemic insult. It is apparent that the cardinal event during anoxia/ischemia is the rapid massive depolarization generally referred to as anoxic depolarization. This depolarization triggers a cascade of effects resulting in cell death. The primary strategy used by the anoxia tolerant turtle and carp brain to avoid anoxic cell damage is to prevent the occurrence of the anoxic depolarization. On the basis of data from the mammal, turtle and carp, there are at least three major variables with respect to the anoxic depolarization: falling levels of intracellular ATP, failure of active ion pumps and the opening of large conductance channels ("megachannels"?). Borrowing from the turtle and carp, the most effective neuroprotective strategy for humans would be either prevention of the anoxic depolarization or more realistically the blocking of the large conductance channels responsible for the anoxic depolarization. Ischemic strokes in humans result from the occlusion of cerebral arterial vessels. The core region of the affected brain tissue cannot be salvaged since cell death would occur too quickly to permit intervention. However, in the penumbral regions where some perfusion is preserved, recurrent anoxic depolarization-like events propagate outward from the border of the core region (Dirnagl *et al.*, 1999; Anderson *et al.*, 2005). These peri-infarct depolarizations would result in expansion of the size of the original core region during the 24 hours following the stroke (Anderson *et al.*, 2005). It has been proposed that these peri-infarct depolarizations have the same underlying mechanism as the anoxic depolarizations which occur in the core region, i.e. the opening of large conductance "megachannels" coupled with active ion pump failure (Anderson *et al.*, 2005). If these large conductance channels could be blocked then peri-infarction depolarizations would be prevented and the size of the infarcted area could be minimized. This is believed to be an attainable neuroprotective intervention.

With respect to future directions, research directed to the following areas would be very useful.

(1) Identifying the large conductance "megachannels" responsible for the anoxic depolarization is clearly a critical piece of information. What are the signal(s) which result in the opening of these channels? What is the mechanism by which a decreased level of intracellular ATP triggers the opening of these channels? Is the link between ATP levels and the opening of large conductance channels a change in the phosphorylation state of channel proteins as has been described for other ion channels and receptors (Storey and Storey, 2004; Bickler *et al.*, 2000; see "Reduction of ATP utilization by the Turtle" section)? Is the opening of these channels linked to failure of active ion pumps such as the Na^+-K^+-ATPase pump and if so what is the linkage? Are similar large conductance "megachannels" present in the brain of the turtle and carp? As discussed in a previous section, there are some data that these channels are smaller in the turtle and carp than in the mammal. However, there have been no known studies specifically directed at identifying the large conductance channels responsible for ischemia induced depolarization in the turtle and carp brain.

Once these channels have been identified in mammals and the turtle and carp, a comparison of their respective (mammals versus turtle/carp) properties would be extremely interesting.

(2) Hypoxia can modulate the function of ion channels in mammalian neurons (Bickler and Donohoe, 2002). Examples include potassium and sodium channels whose function is directly affected by a low oxygen tension. During hypoxia these potassium channels are closed and these sodium channels are opened. Both of these events would result in membrane depolarization. How are these hypoxia induced depolarizations related to the rapid massive (anoxic) depolarization which is due to the opening of large conductance channels? As yet there is still not a clear picture of the role played by specific oxygen-sensing ion channels during the course of anoxia/ischemia. Although a low intracellular ATP concentration appears to trigger the opening of large conductance "megachannels" what role does hypoxia play in this process? Does, in fact, the opening of these large channels require both a low oxygen tension and a low intracellular ATP level?

Oxygen sensing mechanisms in the turtle brain have not been studied to the same extent as in the mammalian brain. As previously discussed, the observation that turtle and carp neurons do not undergo any change in membrane potential during prolonged periods of anoxia strongly suggests that oxygen-sensing potassium and sodium channels comparable to those found in mammalian neurons are not present in the neurons of these vertebrates. However, as discussed in the "Reduction of ATP utilization by the Turtle" section, there is some evidence that hypoxia does reduce ion permeability in the turtle neuron. For example, Ghai and Buck (1999) found that anoxia reduced whole cell conductance in the turtle brain. Hence a low oxygen tension reduces ion permeability in the turtle but there are no data as to which specific ion channels are affected. The best documented example of an effect of hypoxia on channel function in the turtle is the study by Bickler *et al.* (2000; discussed in a previous section). These investigators reported that hypoxia reduces the activity and abundance of NMDA receptors in the turtle neuron. The mechanisms of oxygen sensing in the turtle and carp brain and the effects of hypoxia on specific ion channels are areas that require much more study.

(3) Although anoxic depolarization of the neuron is the critical early event during hypoxia and ischemia, it is hard to believe that this event occurs without the involvement of glial cells given the complex dynamic interplay that occurs between these cells and neurons. Most descriptions of anoxic depolarization treat the neuron as an independent entity, whereas in reality neurons and glia form an interdependent system. Defining the role of glial cells in the anoxic depolarization of mammalian neurons and in the ischemia induced depolarization of turtle/carp neurons is as yet an unexplored area of research.

From the study of vertebrates such as the turtle and carp, we learn that the brain can be designed to withstand severe and prolonged oxygen deprivation without suffering damage. The key design feature is the capacity of the brain of these vertebrates to maintain intracellular ATP levels during a hypoxic state and by so doing prevent the opening of large conductance channels. When these channels open, ion gradients dissipate, the neuronal membrane depolarizes (generally referred to as anoxic depolarization) and a cascade of effects are triggered which propel the neuron down a path towards cell death. The turtle and carp brain maintain ATP levels by an extensive reduction of ATP consumption. Currently it is believed that this reduction in ATP utilization is accomplished by hypoxia induced downregulation of ion channel activity although the evidence for this proposal is still not solid. Clarifying the mechanism by which the turtle and carp brains reduce ATP utilization during hypoxia would be very important since this property is already a feature, although to a lesser extent, of the relatively hypoxia tolerant mammalian neonatal brain. Could this property (i.e. hypoxia induced reduction in ATP utilization) be enhanced in the mammalian brain? As yet there is no biological solution for the problem of ischemic brain damage. The brain of the turtle and the brain of the carp, although extremely hypoxia tolerant, are still sensitive to ischemic damage. However, identifying the large conductance channels responsible for the anoxic depolarization would be an important first step in developing a plausible strategy for neuroprotection during ischemic strokes in humans; the designing of agents to block these channels and hence prevent expansion of the core ischemic area into the penumbral regions. Designing such agents would clearly be facilitated by knowledge of how the properties of these channels in the turtle and carp compare to those in the mammal.

References

Anderson, T.R., Jarvis, C.R., Biedermann, A.J., Molnar, C., Andrew, R.D., 2005. Blocking the anoxic depolarization protects without functional compromise following simulated stroke in cortical brain slices. *J. Neurophys.* 93, 963–979.

Ashcroft, F.M., Gribble, F.M., 1998. Correlating structure and function in ATP-sensitive K+ channels. *Trends. Neurosci.* 21, 288–294.

Auer, R.N., Sutherland, G.R., 2002. Hypoxia and related conditions. In: Graham, D.I., Lantos, P.L. (Eds.), *Greenfield's Neuropathology*, 7th Edn. Arnold (a member of the Hodder Headline Group), London, England, pp. 233–280.

Ballanyi, K., Doutheil, J., Brockhaus, J., 1996. Membrane potentials and microenvironment of rat dorsal vagal cells *in vitro* during energy depletion. *J. Physiol.* 495(3), 769–784.

Ballanyi, K., Kulik, A., 1998. Intracellular Ca^{2+} during metabolic activation of K_{ATP} channels in spontaneously active dorsal vagal neurons in medullary slices. *Eur. J. Neurosci.* 10, 2574–2585.

Ballanyi, K., 2004. Protective role of neuronal K_{ATP} channels in brain hypoxia. *J. Exp. Biol.* 207, 3201–3212.

Barber, P.A., Demchuk, A.M., Hirt, L., Buchan, A.M., 2003. Biochemistry of ischemic stroke. In: Barnett, H.J.M., Bogousslavsky, J., Meldrum, H. (Eds.), *Advances in Neurology, Volume 92, Ischemic Stroke.* Lippincott Williams and Wilkins, Philadelphia, PA, USA, pp. 151–164.

Bickler, P.E., Gallego, S.M., Hansen, B.M., 1993. Developmental changes in intracellular calcium regulation in rat cerebral cortex during hypoxia. *J. Cereb. Blood Flow Metab.* 13, 811–819.

Bickler, P.E., Hansen, B.M., 1994. Causes of calcium accumulation in rat cortical brain slices during hypoxia and ischemia: role of ion channels and membrane damage. *Brain Res.* 665, 269–276.

Bickler, P.E., Buck, L.T., 1998. Adaptations of vertebrate neurons to hypoxia and anoxia: maintaining critical Ca^{2+} concentrations. *J. Exp. Biol.* 201, 1141–1152.

Bickler, P.E., Hansen, B.M., 1998. Hypoxia-tolerant neonatal CA1 neurons: relationship of survival to evoked glutamate release and glutamate receptor-mediated calcium changes in hippocampal slices. *Dev. Brain Res.* 106, 57–69.

Bickler, P.E., Donohoe, P.H., Buck, L.T., 2000. Hypoxia-induced silencing of NMDA receptors in turtle neurons. *J. Neurosci.* 20, 3522–3528.

Bickler, P.E., Donohoe, P.H., 2002. Adaptive responses of vertebrate neurons to hypoxia. *J. Exp. Biol.* 205, 3579–3586.

Bickler, P.E., Fahlman, C.S., Taylor, D.M., 2003. Oxygen sensitivity of NMDA receptors: relationship to NR2 subunit composition and hypoxia tolerance of neonatal neurons. *Neuroscience* 118, 25–35.

Binstock, L., Lecar, H., 1969. Ammonium ion currents in the squid giant axon. *J. Gen. Physiol.* 53, 342–361.

Brown, M.M., Wade, J.P.H., Marshall, J., 1985. Fundamental importance of arterial oxygen content in the regulation of cerebral blood flow in man. *Brain* 108, 81–93.

Buxton, R.B., Frank, L.R., 1997. A model for the coupling between cerebral blood flow and oxygen metabolism during neural stimulation. *J. Cereb. Blood Flow Metab.* 17, 64–72.

Buxton, R.B., 2001. The elusive initial dip. *NeuroImage* 13, 953–958.

Chih, C.-P., Feng, Z.-C., Rosenthal, M., Lutz, P.L., Sick, T.J., 1989a. Energy metabolism, ion homeostasis, and evoked potentials in anoxic turtle brain. *Am. J. Physiol.* 257, R854–R860.

Chih, C.-P., Rosenthal, M., Sick, T.J., 1989b. Ion leakage is reduced during anoxia in turtle brain: a potential survival strategy. *Am. J. Physiol.* 257, R1562–R1564.

Cooper, A.J.L., Plum, F., 1987. Biochemistry and physiology of brain ammonia. *Physiol. Rev.* 67, 440–519.

Cummins, T.R., Donnelly, D.F., Haddad, G.G., 1991. Effect of metabolic inhibition on the excitability of isolated hippocampal CA1 neurons: developmental aspects. *J. Neurophys.* 66, 1471–1482.

Diamond, J.S., 2006. Astrocytes put down the broom and pick up the baton. *Cell* 125, 639–641.

Dirnagl, U., Iadecola, C., Moskowitz, M.A., 1999. Pathobiology of ischemic stroke: an integrated view. *Trends Neurosci.* 22, 391–397.

Doll, C.J., Hochachka, P.W., Reiner, P.B., 1991. Effects of anoxia and metabolic arrest on turtle and rat cortical neurons. *Am. J. Physiol.* 260, R747–R755.

Doll, C.J., Hochachka, P.W., Reiner, P.B., 1993. Reduced ionic conductance in turtle brain. *Am. J. Physiol.* 265, R929–R933.

Douglas, H.A., Yang, G., Andrew, R., Kirov, S.A., 2006. Two-photon microscopical study of pyramidal neurons protected from ischemic stress by dibucaine. *Soc. Neurosci.* (abstract, in press).

Duffy, T.E., Kohle, S.J., Vannucci, R.C., 1975. Carbohydrate and energy metabolism in perinatal rat brain: relation to survival in anoxia. *J. Neurochem.* 24, 271–276.

Faraci, F.M., 1991. Adaptations to hypoxia in birds: how to fly high. *Ann, Rev. Physiol.* 53, 59–70.

Farahani, R., Pina-Benabou, M.H., Kyrozis, A., Siddiq, A., Barradas, P.C., Chiu, F-C., Cavalcante, L.A., Lai, J.C.K., Stanton, P.K., Rozental, R., 2005. Alterations in metabolism and gap junction expression may determine the role of astrocytes as "good Samaritans" or executioners. *Glia* 50, 351–361.

Fried, E., Amorim, P., Chambers, G., Cottrell, J.E., Kass, I.S., 1995. The importance of sodium for anoxic transmission damage in rat hippocampal slices: mechanisms of protection by lidocaine. *J. Physiol.* 489(2), 557–565.

Ghai, H.S., Buck, L.T., 1999. Acute reduction in whole cell conductance in anoxic turtle brain. *Am. J. Physiol.* 277, R887–R893.

Gjedde, A., Marrett, S., 2001. Glycolysis in neurons, not astrocytes, delays oxidative metabolism of human visual cortex during sustained checkerboard stimulation *in vivo. J. Cereb. Blood Flow Metab.* 21, 1384–1392.

Gjedde, A., Marrett, S., Vafaee, M., 2002. Oxidative and non-oxidative metabolism of excited neurons and astrocytes. *J. Cereb. Blood Flow Metab.* 22, 1–14.

Groebe, K., Thews, G., 1992. Basic mechanisms of diffusive and diffusion-related oxygen transport in biological systems: a review. *Adv. Exp. Med. Biol.* 317, 21–33.

Haddad, G.G., Donnelly, D.F., 1990. O_2 deprivation induces a major depolarization in brain stem neurons in the adult but not in the neonatal rat. *J. Physiol.* 429, 411–428.

Haddad, G.G., Jiang, C., 1993a. O_2 deprivation in the central nervous system: on mechanisms of neuronal response, differential sensitivity and injury. *Prog. Neurobiol.* 40, 277–318.

Haddad, G.G., Jiang, C., 1993b. Mechanisms of anoxia-induced depolarization in brainstem neurons: *in vitro* current and voltage clamp studies in the adult rat. *Brain Res.* 625, 261–268.

Hammarstrom, A.K.M., Gage, P.W., 1998. Inhibition of oxidative metabolism increases persistent sodium current in rat CA1 hippocampal neurons. *J. Physiol.* 510(3), 735–741.

Hammarstrom, A.K.M., Gage, P.W., 2000. Oxygen-sensing persistent sodium channels in rat hippocampus. *J. Physiol.* 529(1), 107–118.

Hammarstrom, A.K.M., Gage, P.W., 2002. Hypoxia and persistent sodium current. *Eur. Biophys. J.* 31, 323–330.

Hansen, A.J., 1977. Extracellular potassium concentration in juvenile and adult rat brain cortex during anoxia. *Acta Physiol. Scand.* 99, 412–420.

Hansen, A.J., 1985. Effect of anoxia on ion distribution in the brain. *Physiol. Rev.* 65, 101–148.

Hochachka, P.W., Lutz, P.L., 2001. Mechanism, origin, and evolution of anoxia tolerance in animals. *Comp. Biochem. Physiol. Part B* 130, 435–459.

Hoge, R.D., Atkinson, J., Gill, B., Crelier, G.R., Marrett, S., Pike, G.B., 1999. Linear coupling between cerebral blood flow and oxygen consumption in activated human cortex. *Proc. Natl. Acad. Sci. USA* 96, 9403–9408.

Jarvis, C.R., Anderson, T.R., Andrew, R.D., 2001. Anoxia depolarization mediates acute damage independent of glutamate in neocortical brain slices. *Cereb. Cortex* 11, 249–259.

Jiang, C., Haddad, G.G., 1994. A direct mechanism for sensing low oxygen levels by central neurons. *Proc. Natl. Acad. Sci. USA* 91, 7198–7201.

Johansson, D., Nilsson, G.E., 1995. Role of energy status, K_{ATP} channels and channel arrest in fish brain K^+ gradient dissipation during anoxia. *J. Exp. Biol.* 198, 2575–2580.

Jones, M., Berwick, J., Johnston, D., Mayhew, J., 2001. Concurrent optical imaging spectroscopy and laser-doppler flowmetry: the relationship between blood flow, oxygenation, and volume in rodent barrel cortex. *Neuroimage* 13, 1002–1015.

Kety, S.S., Schmidt, C.F., 1948. The effects of altered arterial tensions of carbon dioxide and oxygen on cerebral blood flow and cerebral oxygen consumption of normal young men. *J. Clin. Invest.* 27, 484–492.

Kulik, A., Trapp, S., Ballanyi, K., 2000. Ischemia but not anoxia evokes vesicular and Ca^{2+}-independent glutamate release in the dorsal vagal complex in vitro. *J. Neurophys.* 83, 2905–2915.

LaManna, J.C., Vendel, L.M., Farrell, R.M., 1992. Brain adaptation to chronic hypobaric hypoxia in rats. *J. Appl. Physiol.* 72, 2238–2243.

LaManna, J.C., Chavez, J.C., Pichiule, P., 2004. Structural and functional adaptation to hypoxia in the rat brain. *J. Exp. Biol.* 207, 3163–3169.

Liang, Y.-H., Liu, X.-Z., Liu, S.-H., Lu, G.-Y., 2001. The structure of greylag goose oxy haemoglobin: the roles of four mutations compared with barheaded goose haemoglobin. *Acta Crystallogr. D Biol. Crystallogr.* 57, 1850–1856.

Lindauer, U., Gethmann, J., Kuhl, M., Kohl-Bareis, M., Dirnagl, U., 2003. Neuronal activity-induced changes of local cerebral microvascular blood oxygenation in the rat: effect of systemic hyperoxia or hypoxia. *Brain Res.* 975, 135–140.

Lipton, P., 1999. Ischemic cell death in brain neurons. *Physiol. Rev.* 79, 1431–1568.

Lutz, P.L., Milton, S.L., 2004. Negotiating brain anoxia survival in the turtle. *J. Exp. Biol.* 207, 3141–3147.

Nilsson, G.E., 2001. Surviving anoxia with the brain turned on. *News Physiol. Sci.* 16, 217–221.

Nilsson, G.E., Renshaw, G.M.C., 2004. Hypoxic survival strategies in two fishes: extreme anoxia tolerance in the North European crucian carp and natural hypoxic preconditioning in a coral-reef shark. *J. Exp. Biol.* 207, 3131–3139.

Panatier, A., Theodosis, D.T., Mothet, J.-P., Touquet, B., Pollegioni, L., Poulain, D.A., Oliet, S.H.R., 2006. Glia-derived D-serine controls NMDA receptor activity and synaptic memory. *Cell* 125, 775–784.

Pearigen, P., Gwinn, R., Simon, R.P., 1996. The effects *in vivo* of hypoxia on brain injury. *Brain Res.* 725, 184–191.

Plant, L.D., Kemp, P.J., Peers, C., Henderson, Z., Pearson, H.A., 2002. Hypoxic depolarization of cerebellar granule neurons by specific inhibition of TASK-1. *Stroke* 33, 2324–2328.

Scott, G.R., Milsom, W.K., 2006. Flying high: a theoretical analysis of the factors limiting exercise performance in birds at altitudes. *Respir. Physiol. Neurobiol.* (in press).

Shimojyo, S., Scheinberg, P., Kogure, K., Reinmuth, O.M., 1968. The effects of graded hypoxia upon transient cerebral blood flow and oxygen consumption. *Neurology* 18, 127–133.

Sick, T.J., Lutz, P.L., LaManna, J.C., Rosenthal, M., 1982a. Comparative brain oxygenation and mitochondrial redox activity in turtles and rats. *J. Appl. Physiol. Respir. Environ. Exerc. Physiol.* 53, 1354–1359.

Sick, T.J., Rosenthal, M., LaManna, J.C., Lutz, P.L., 1982b. Brain potassium ion homeostasis, anoxia, and metabolic inhibition in turtles and rats. *Am. J. Physiol.* 243, R281–R288.

Soengas, J.L., Aldegunde, M., 2002. Energy metabolism of fish brain. *Comp. Biochem. Physiol. Part B* 131, 271–296.

Storey, K.B., Storey, J.M., 2004. Metabolic rate depression in animals: transcriptional and translational control. *Biol. Rev.* 79, 207–233.

Suarez, R.K., Doll, C.J., Buie, A.E., West, T.G., Funk, G.D., Hochachka, P.W., 1989. Turtles and rats: a biochemical comparison of anoxia–tolerant and anoxia-sensitive brains. *Am. J. Physiol.* 257, R1083–1088.

Tanaka, E., Yamamoto, S., Kudo, Y., Mihara, S., Higashi, H., 1997. Mechanisms underlying the rapid depolarization produced by deprivation of oxygen and glucose in rat hippocampal CA1 neurons *in vitro. J. Neurophysiol.* 78, 891–902.

Tanaka, E., Yamamoto, S., Inokuchi, H., Isagai, T., Higashi, H., 1999. Membrane dysfunction induced by *in vitro* ischemia in rat hippocampal CA1 pyramidal neurons. *J. Neurophys.* 81, 1872–1880.

Thompson, R.J., Zhou, N., MacVicar, B.A., 2006. Ischemia opens neuronal gap junction hemichannels. *Science* 312, 924–927.

Warren, D.E., Reese, S.A., Jackson, D.C., 2006. Tissue glycogen and extracellular buffering limit the survival of red-eared slider turtles during anoxic submergence at 3°C. *Physiol. Biochem. Zool.* 79, 736–744.

Xia, Y., Jiang, C., Haddad, G.G., 1992. Oxidative and glycolytic pathways in rat (newborn and adult) and turtle brain: role during anoxia. *Am. J. Physiol.* 262, R595–R603.

Yamashita, A., Makita, K., Kuroiwa, T., Tanaka, K., 2006. Glutamate transporters GLAST and EAAT4 regulate post-ischemic Purkinje cell death: an *in vivo* study using a cardiac arrest model in mice lacking GLAST or EAAT4. *Neurosci. Res.* 55, 264–270.

Young, R.S.K., During, M.J., Donnelly, D.F., Aquila, W.J., Perry, V.L., Haddad, G.G., 1993. Effects of anoxia on excitatory amino acids in brain slices of rats and turtles: *in vitro* microdialysis. *Am. J. Physiol.* 264, R716–R719.

Epilogue

Evolution can be considered a natural experimental process in which numerous animal design features are tested. The premise of this book is that adaptations selected out by specific environmental pressures actually represent biological solutions to certain human clinical disorders. Since the natural animal model is living proof that a biological solution to a given clinical disorder exists, then understanding the biochemical/physiological characteristics of that natural animal model should clarify the features of that solution. Before proceeding, the distinction between the natural and the traditional animal model should be emphasized since this distinction is a cornerstone of the premise of this book. This distinction has been discussed in the introductory chapter. For animals used in medical research, the usual protocol is to induce a disease in the animal, assume the disease (in that animal) is similar to that observed in humans and then do appropriate experiments using this "traditional" animal model. The important point is that this animal is not in its natural state and treatment strategies developed experimentally through the use of such an animal model are "external" to that animal. That is to say, the treatment strategy has been imposed from the outside. A natural animal model does not have a disease but by definition has adapted (i.e. possesses a biological solution) to biochemical/physiological characteristics which in the human would have pathological consequences. In essence, this animal has developed a "treatment strategy" through the natural experimental process of evolution for what would be a disease in humans. This "treatment strategy" is "internal" to that animal and as noted in the introductory chapter this treatment strategy has evolved within the multifunctional framework that is characteristic of the design of animal systems.

The topics comprising the six chapters were chosen arbitrarily based upon my familiarity with these clinical disorders as well as upon knowledge regarding suitable natural animal models. In each chapter, I have reviewed the clinical disorder with as much detail as I thought necessary to allow for comparison with the natural animal model. No doubt, relevant references have probably been overlooked and in so doing the discussion of the clinical disorder may appear biased. Considering the book as a whole, the following few general observations are made.

(1) Although each chapter can be read independently of the others, this implies a compartmentalization which is really unrealistic. It is clear that there is overlap. For example, the adaptations that the bear has evolved to minimize the degree of osteoporosis and muscle atrophy resulting from a prolonged period of inactivity (hibernation) are clearly also part of the set of adaptations that the bear has evolved to handle a prolonged period of functional renal failure. The presentation of these clinical disorders in different chapters (chronic renal failure in Chapter 2 and disuse osteoporosis and disuse muscle atrophy in Chapter 4) gives the illusion that the bear's adaptations (biological solutions) to these problems are separate. They are not and the key adaptations are probably the bear's ability to recycle nitrogenous wastes and maintain a normal protein synthetic rate during its prolonged dormant period. The mechanisms underlying the adverse effects of increased brain ammonia concentrations on neuronal function (Chapter 5) appear to be quite similar to the mechanisms underlying the neurotoxic effects of hypoxia/ischemia (Chapter 6). Hence vertebrates which are extremely ammonia tolerant are probably also hypoxia tolerant, and similarly vertebrates which are extremely hypoxia tolerant are probably ammonia tolerant. However, the possible linkage between ammonia and hypoxia tolerance has not been investigated although this would appear to be a research area well worth exploring.

(2) The biological solution present within a given natural animal model does not necessarily represent a unique set of biochemical/physiological characteristics. These biochemical/physiological characteristics are probably present in most vertebrates. In general, the natural animal model builds on this existing set of biochemical/physiological characteristics. Several examples supporting this generalization are as follows. Neurons of the mammalian neonate and the turtle have developed similar strategies for avoiding hypoxic damage. However, the extreme hypoxia tolerance of the

turtle brain has been attained by extending and "fine tuning" these strategies (Chapter 6). Similar to the bear, most mammals recycle urea through the gastrointestinal tract and use a variable fraction of the salvaged urea nitrogen for the synthesis of proteins. The bear is able to withstand a prolonged period of functional renal failure because this animal is much more efficient in recycling urea and using essentially all of the nitrogen for protein synthesis (Chapter 2). In essence, the natural animal model appears to have extended and "fine tuned" existing biochemical/physiological characteristics. This generalization is also supported by the data on mammalian hibernators. As pointed out by Carey *et al.* (2003) and Boyer and Barnes (1999), the widespread distribution of mammalian species that hibernate is consistent with the hypothesis that the genes required to specify the hibernating phenotype are common among the genomes of all mammals.

(3) The natural animal model's biological solution to a clinical disorder is based upon a strategy of early intervention. The following examples illustrate this point. The turtle avoids hypoxic brain damage by preventing the occurrence of anoxic depolarization in the neuron (Chapter 6). Although the coronary arteries in the fish develop intimal thickenings, these thickenings do not progress to form the complex lipid-laden lesions observed in humans (Chapter 3). Although still speculative, it appears that the bird avoids tissue damage secondary to chronic high (by human standards) blood glucose concentrations by controlling the influx of glucose into cells (Chapter 1).

In addition, there is a message in this "early intervention" strategy with respect to our own research directions. In the case of hypoxic/ischemic brain damage, much research has been directed at "blocking" secondary events that follow the anoxic depolarization, such as the elevation of brain extracellular glutamate concentration and activation of NMDA receptors. Much less research has been devoted to defining the mechanism underlying the anoxic depolarization and how this event could be prevented. In the case of human atherosclerotic vascular disease, current therapeutic strategies involve the use of lipid lowering agents to prevent additional vascular lesions and to promote regression of existing lesions. The natural animal model suggests that perhaps we should be trying to modify the microenvironment of the arterial intima such as to prevent the development of complex lesions in the first place.

(4) As one reads this book, it becomes apparent that the volume of published data relevant to the biological solutions of the various natural animal models is in general much less than the volume of published

data with respect to the various clinical disorders. For example, there are only a few published papers dealing with the ability of the hibernating bear to avoid disuse muscle atrophy (Chapter 4). I believe this paucity of data on natural animal models has occurred because natural animal models represent an interface area of research. Several examples can be used to illustrate this point. Certain species of fish have been documented to develop coronary artery wall thickenings. This finding is clearly important from a fish physiology perspective, but performing experiments on these fish which relate this finding to the pathogenesis of atherosclerotic vascular disease in mammals (humans) appears to be outside the mainstream of research for both fish physiologists/biochemists and for medical scientists working in cardiovascular research. The hibernating bear contends with a prolonged period of functional renal failure without suffering any adverse consequences. However, the hibernating bear has not been embraced by scientists working in the field of nephrology as a worthy model for studying chronic renal failure. A similar story is true for the bird as a natural animal model for studying diabetes mellitus. The turtle appears to be an exception to this trend. Medical scientists and biologists have recognized that the turtle is a useful model for studying the pathogenesis of hypoxic brain damage since this vertebrate can withstand prolonged anoxia without experiencing any brain injury. However, the turtle brain is not resistant to ischemia and when exposed to an ischemic insult the turtle neuron undergoes anoxic depolarization. This particular important event (i.e. anoxic depolarization of ischemic turtle brain) has not been studied at all.

Because of the limited data with respect to the various natural animal models, a number of basic questions remain unanswered. These questions have been discussed in the individual chapters, but only a few are listed here. Glucose transporters have been identified in birds but have not been characterized to the same extent as glucose transporters in mammals. Hence, my speculation that in birds glucose transporters regulate the influx of glucose into cells such as to prevent intracellular "toxic" metabolic processes from inducing tissue damage must remain just that — a hypothesis (Chapter 1). The bear most certainly has urea transporters like other mammals yet the presence of urea transporters has not actually been documented in this mammal. The hypothesis that urea transporters in the liver, gastrointestinal tract, kidney and urinary bladder of the hibernating bear are somehow regulated to efficiently recycle almost 100% of urea must remain a speculation until there are data on urea transporters in this animal

(Chapter 2). The biology of vascular smooth muscle and endothelial cells in the fish (e.g. salmon, trout) has barely been explored and so one can only hypothesize that the microenvironment of the fish vascular wall as opposed to that of the mammal, is in some way resistant to the development of complex lipid laden atherosclerotic lesions (Chapter 3). Although fish in general and some species in particular are ammonia tolerant, the number of published studies directly examining what adaptations of the fish brain are responsible for this tolerance are few in number (Chapter 5). The large conductance channels, the opening of which is responsible for the anoxic depolarization in hypoxic/ischemic mammalian neurons have not yet been identified. Although the turtle and carp brain are extremely anoxia tolerant, they are not tolerant of ischemia. The neurons of both of these vertebrates undergo depolarization when subjected to an ischemic insult. However this event, the depolarization of ischemic turtle and carp brain, has not been studied to any extent. Do the neurons of the turtle and carp have large conductance channels which remain closed during anoxia but open during ischemia? If so how do these channels compare to those of the mammalian neuron? There are no data to answer these questions (Chapter 6).

(5) Finally, as one reads through all of these chapters, one is struck by the apparent paradox between the rich diversity of life forms and the marked conservatism of body system designs. It is this conservatism in the design of body systems that opens up the possibility of transferring the biological solutions of natural animal models to humans.

Finally, I wish to end on a personal note with a comment on why I have written this book since doing so has been both a rewarding and challenging experience. The study of natural animal models in my estimation represents a realistic and powerful tool for exploring the pathogenesis of different human diseases and for developing workable treatment strategies. This approach would complement existing research methodologies. As I have already stated, the natural animal model is living proof that a biological solution to a given clinical disorder exists. The study of natural animal models is an area of research that bridges biology and medical science. Each chapter in this book presents a different clinical disorder and an appropriate natural animal model as an illustration of how that natural animal model has solved that particular clinical disorder. This book is not meant to be a primer on the various clinical disorders nor a primer on comparative physiology. I will have accomplished my goal if I succeed in convincing at least one researcher to consider using this approach. Since I am a physician by training, I would find it very exciting if this book encouraged for example, cardiologists to consider the fish and nephrologists the bear as worthy models for studying the problems of atherosclerotic vascular disease and chronic renal failure, respectively.

Michael A. Singer
Professor Emeritus
Faculty of Medicine
Queen's University
Kingston, Ontario, Canada
K7L 3N6

References

Boyer, B.B., Barnes, B.M., 1999. Molecular and metabolic aspects of mammalian hibernation. *BioScience* 49, 713–724.

Carey, H.V., Andrews, M.T., Martin, S.L., 2003. Mammalian hibernation: cellular and molecular responses to depressed metabolism and low temperature. *Physiol. Rev.* 83, 1153–1181.

Index